A New Plateau

SUSTAINING THE LANDS AND PEOPLES OF CANYON COUNTRY

A New Plateau

SUSTAINING THE LANDS AND PEOPLES OF CANYON COUNTRY

EDITORS:

Peter Friederici and Rose Houk

SENIOR PHOTOGRAPHER:

Tony Marinella

ART DIRECTION AND DESIGN:

Brett Olson

INTRODUCTION BY:

Gary Paul Nabhan

A PROJECT OF THE

Center for Sustainable Environments, Northern Arizona University

with assistance from the Museum of Northern Arizona

PUBLISHED BY

Renewing the Countryside

A New Plateau: Sustaining the Lands and Peoples of Canyon Country

Editors	Peter Friederici Rose Houk
Senior Photographer	Tony Marinella
Art Direction	Brett Olson

Writers

Roger Clark	Rosemary Logan
Peter Friederici	Gary Paul Nabhan
Rachel Turiel Hinds	Sue Norris
Rose Houk	Tony Norris
Charles E. Jones	Roberto Nutlouis
Patty Kohany	Ashley Rood
Susan Lamb	David Seibert
Charlie Laurel	Tim Swinehart

Photographers and Illustrators

George Andrejko	Mary Layman
Tom Bean	Rosemary Logan
Dan Boone	Tony Marinella
Greg Bryan	Michael Mertz
Chalk Hill, Inc.	Tony Norris
Teresa DeKoker	Roberto Nutlouis
David Edwards	Ashley Rood
Peter Friederici	John Running
Rick Gilliam	David Seibert
Carol Snyder Halberstadt	Eric Swanson
Bruce Hucko	Patty West
Charlie Laurel	

Printer	Friesens, Canada
Paper	100 lb. Reincarnation Matte, 100% recycled, 50% post consumer waste, New Leaf Paper, San Francisco
Cover Photography	Cover background: San Francisco Peaks at dawn, Tony Marinella Back cover: Sunset over Black Mesa, Courtesy Black Mesa Weavers for Life and Land, ©1999, Carol Snyder Halberstadt

ISBN-0-9713391-5-5 (hardcover); ISBN-0-9713391-6-3 (paperback)
Library of Congress Control Number: 2004097149

First Printing

*Dedicated to the
people of the Southwest
who have "dug in" and
planted the seeds of
sustainability.*

TABLE OF CONTENTS

Section Two: Ranching

Section Three: Wildcrafting

Section Four: Sustainable Building and Energy

SECTION FIVE: REGIONAL FOOD HERITAGE

FOREWORD

By Robert G. Breunig

Director, Museum of Northern Arizona

For at least ten thousand years, until the arrival of railroads in the late 1880s, the people of the Colorado Plateau – or Canyon Country – primarily derived their sustenance from the natural resources of the land they inhabited. Earlier cultures lived here by collecting and growing their own food, and by hunting. The water they used was provided by springs and a few permanent streams and rivers. Demands quite different from those of today challenged their lives, requiring a deep knowledge and intimate understanding of the land's resources and natural systems.

People have entered this millennium in a vastly different situation. With increased use of the railroad, highways, pipe-lines, and other technological developments, human dependence on local resources has declined radically. Now almost everything we eat, wear, and use on the Colorado Plateau comes from other places. A byproduct of this transformation has been a progressive loss of intimacy with the land we call home. This is regrettable; human life is richer and more inherently rewarding when people feel a deep personal connection with the place in which they live.

In today's Canyon Country, a technological, import-reliant "life support system" provides for most of our material needs. We are beginning to learn that this system has inherent constraints, and in the long term it may fail to provide us with reliable sources of sustenance. Already we know that an extended period of drought threatens the reservoir system we have come to depend on for our water and energy supplies. Rising fuel prices remind us that petroleum is a finite resource that soon will reach its peak of global production, bringing still higher prices and, potentially, political and social disruption. In addition, rising rates of disease, associated with the radical transformation of our food supplies and eating habits, remind us how far we have strayed from the regional foods we once collected and grew with our own hands.

A New Plateau chronicles the achievements of an inspired group of Canyon Country people who are countering these trends by asserting a new kind of citizenship on the plateau – a citizenship that extends beyond the political realm to root itself in a deep respect for, and a renewed reliance on, the nature of their region. These people seek to understand this region's natural limitations and uncover sustainable ways to harvest and capture its resources. They are cultural pioneers, pointing us in new, more sustainable directions.

These new pioneers are making lifetime commitments to the health and quality of the region while celebrating its diverse landscapes, languages, lifestyles, foods, and flavors. In their efforts they are drawing from ancient cultural traditions as well as from new technologies. Their achievements hold promise for the ecological health and future livability of the Colorado Plateau region. *A New Plateau* is an inspired background for their dialogue, one that honors ancient principles and understandings while charting creative directions for achieving a sustainable living on this land.

INTRODUCTION

By Gary Paul Nabhan
Director, Center for Sustainable Environments
Northern Arizona University

For some of the twelve million people who visit the Four Corners Canyon Country every year, the experience of finding thousand-year-old corn cobs in a cliff dwelling is an amazing connection with the past – and a revelation. How, they wonder, did a crop more often associated with the lush fields of the Midwest turn up in such an arid land? How did the ancients survive amid the vast expanses of rock and sand? Of course, the answers to these questions can be learned at any number of Canyon Country visitors' centers: that corn has been grown on the plateaus and in the canyons of Arizona, Utah, Colorado, and New Mexico for nearly four thousand years – and that this tradition continues today.

The same visitors, indeed, find themselves amazed that so many different cultures have successfully adapted to this wind-swept, water-limited landscape. They are astonished to learn that some bean and corn fields at Zuni have been cultivated year in and year out for centuries without depleting the fertility of the soil. They are awed by the thick-walled rock and mud pueblos of Hopi and Acoma, where contemporary inhabitants comfortably dwell in the very same rooms that their great-great grandparents built well over a century ago.

Sooner or later, visitors realize that it is not just the Ancient Ones who adapted their lives to the seasons, soils, droughts,

and dramatic topography of Canyon Country. Rather, Native, Hispanic, Anglo, and Basque Americans are still doing so today. In a very real sense, the Canyon Country of the Four Corners states remains a living laboratory of grassroots experiments to advance sustainability – that is, to gain food, fiber, fuel, and shelter in a manner that enriches rather than limits possibilities for future generations. To be sure, some among the region's inhabitants have overharvested forests or drained groundwater reserves, but others have always sought to live, as Henry David Thoreau envisioned, by "meeting the expectations of the land." The southwestern United States has been home to certain large-scale experiments to advance sustainable technologies – from renewable energy generation at Sandia Labs in New Mexico to Paolo Soleri's "arcology" experiments at the village of Arcosanti near Cordes Junction, Arizona. But for the most part, the successful efforts in advancing sustainability have come up from the grassroots, at a scale appropriate to the canyons and pueblos themselves.

One can travel from national park to national park in the Southwest and never recognize that within just a few miles of Route 66 or the Coronado Trail a whole host of remarkable experiments are in progress. Families, collectives, and communities are learning how to grow their food, frame their shelters, and generate their energy with the natural resources and cultural knowledge found at their doorsteps. They are fueling their efforts with energy from the wind and the sun, and with small-diameter timber thinned to reduce fire hazards in the ever-changing forests and woodlands that surround them. They are rediscovering the remarkable environmental adaptations of native seeds, heirloom fruits, and hardy breeds of livestock. They are wildcrafting native medicines, local clays, dyes, and fiber plants for their tinctures, pottery, weavings, and basketry.

Of course, cynics may claim that these efforts to live sustainably are laudable but not very lucrative. Ironically, many of the pioneering innovators in sustainable farming and ranching have no time to listen to such critiques. They are laughing all the way to the bank, thanks to the growing markets for environmentally friendly products that celebrate the region's distinctive multicultural heritage. National and global market trends reinforce this idea – various "green" products are already valued at more than $110 billion a year in the United States. A few examples from Canyon Country indicate that this region is a leader in the movement.

Natural foods from sustainably managed farms and ranches. From 1998 to 2003, sales in the multibillion-dollar organic food industry grew 22 percent annually in the U.S. Natural and organic foods are now found in about three-quarters of all grocery stores in the U.S., and are no longer considered affordable only to elite or health-conscious consumers. In Flagstaff, Arizona, alone, the purchase and consumption of locally produced natural foods have increased twenty-fold in just three years, thanks in large

part to the establishment of a local community market, a community-supported agriculture project, and a community wild foraging project.

Renewable energy from the wind and sun. Globally, the growth rate in energy generation from the wind and sun is twice that of natural gas, coal, and hydropower; the International Energy Agency predicts a 3.3 percent annual growth rate in renewable energy use in coming years. In 2002 alone, $7.3 billion of global wind power technology was installed in some 110 countries. Home-scale wind and solar energy generation has also gained a significant foothold within Canyon Country, thanks to a few small, regionally focused businesses. This scale of clean energy generation is especially needed on Native American reservations, where more than half of all homes remain off the grid.

Certified wood from sustainably harvested forests. Consumer demand for certified wood products is growing so rapidly that nearly 2.5 percent of the world's forests are producing certified lumber, and two of the three largest retail suppliers of wood in the world have pledged to acquire all their timber materials from sustainably managed forests. The Forest Stewardship Council has actively assisted hundreds of communities across the country in improving and certifying the management practices of their forests and woodlands, including those of the White Mountain Apache in Arizona. Traditional hogans made from small-diameter pines thinned from fire-prone forests are now being marketed on the Navajo Nation, where tens of thousands of additional housing units will be needed over the next decade.

Contemporary inhabitants of Canyon Country have shown clearly that they are willing to pay more for regionally produced products that leave a cleaner, healthier environment. In Arizona, a statewide survey indicated that nearly half of all consumers are willing to pay greater than current market prices to obtain locally produced fruits, vegetables, meat, and dairy products. They value the nutritional quality and freshness of such foods, especially if they have been grown and transported by environmentally friendly means.

The stories here are about real people living in the pueblos, ranchlands, and small towns of Canyon Country who are renewing the vitality of their communities through the careful use of natural and cultural resources. The individuals and communities profiled are by no means the only innovators – or the only keepers of time-tried traditions – advancing sustainability in our region. Instead, they represent a much larger cadre of inventive, caring individuals who are choosing to fit their lives into the unique environments of Canyon Country, rather than attempting to remake their environments at the expense of groundwater reserves, artesian springs, endemic plants, and endangered wildlife. They are using, in Wendell Berry's words, nature as a measure of how fitting and lasting their own contributions may ultimately be. If their homes, fields, and orchards last as long as those around the ancient cliff dwellings of Canyon Country, perhaps future generations will marvel at the levels of sustainability that our own contemporaries were able to achieve.

This book is not only meant to place the spotlight on a select few, but also to inspire all readers and all residents of the region to give more care and consideration to the living treasures in our own backyards, and to pass them on to future generations without depleting the aquifers, soils, diverse wildlife habitats, and cultural traditions characteristic of this sun-dried region. We hope that the number of stories about projects that sustain rather than squander the uniqueness of Canyon Country grows exponentially, so that in another decade we will need a massive encyclopedia rather than a modest book to capture all of them.

We have been overjoyed by our adventures in tracking down the farmers, wildcrafters, builders, and wind-catchers whom you are about to meet. We hope that as you travel the backroads and byways of Canyon Country, you remember that around the next corner – down the next switchback of the Moqui Steps or the Mormon Trail – you may find someone who is diligently tending to the land, harvesting a healing herb, or herding an ancient breed of sheep. Pull over on the roadside, shout out a greeting, and join them. With them, you will be helping to sow the seeds of the future.

SECTION ONE
Farming and Market Gardening

Food is basic to life, yet we so often take for granted what we eat. We rarely stop to question where our food was grown, how far it has to come to reach us, or how much fuel and fertilizer it required. For those who farm and garden organically, though, this most basic necessity is never something to take for granted.

In this section you will read stories of people who know exactly where their food comes from and who are up to the challenges of growing it in Canyon Country. The usual unpre-dictabilities of farming – insects, storms, disease – prevail all too often here, along with some region-specific twists. Much of this is high, dry country. Water is always a critical commodity, and its delivery to crops in the right amounts at the right time becomes a fine art. The region's higher elevations mean a shorter growing season, and farmers and gardeners must employ an array of tricks to extend that season.

Some of the people featured here started with dreams of getting back to the land, but once they found their perfect Eden they learned in a hurry that a good dose of common sense and cold economics was necessary. Flexibility is the key to success, along with a readiness to abandon years of hard work if a better way to do something presents itself. Says Cory Rade, "It's one grand experiment, really; every year that's what we do."

Canyon Country farmers have also discovered that organic agriculture is a labor-intensive endeavor that requires constant attention. For the Masayesva family on the Hopi mesas, it means walking to the fields at dawn, weeding, chasing away birds, and keeping the springs cleaned out. For John Sutcliffe, it means carefully pruning each of thousands of grapevines at his vineyard in a Colorado canyon.

But even after years of hard work and hard-won lessons, the dreams of these people are still alive. They take seriously their role as stewards of the land, as keepers of tradition, as sustainers of family and community. Bob Kauer, who owns a farm outside Durango, Colorado, always wanted to have a big garden that could be worked and tended by a broad array of people. Now he's living his dream by providing an acre of land with water where townsfolk can grow their own vegetables. His example, like those of others here, shows that healthy food, healthy land, and healthy communities go hand in hand.

In the Shadow of the Mountains
Making a Living at Stone Free Farm

Arriola, Colorado · By Rachel Turiel Hinds

*Rosie Carter and
Chuck Barry*

In winter, farmers sink into worn couches, sip tea, and tell stories about summers past. After a decade of farming the high desert outside of Cortez, Colorado, Chuck Barry and Rosie Carter have stories. This winter the story begins with water: for the first time in many years, their three-acre farm has been draped in a deep coat of snow.

Last summer the stories were of hail – violent downpours that ruined the pepper and tomato crops. The summer before that it was severe drought – the driest year on Colorado record. While many other farmers put their tools away for good, the drought brought lasting wisdom to Stone Free Farm, shaping the way the farm has evolved.

"The drought taught us what works for a small farm business," Rosie shares from their just-big-enough-for-two, original homestead house. "We realized it's not worth it financially to put water into crops that take all season to grow, like winter squash or cabbage. Instead, we plant intensively, a new row each week, with high crop turnover. Every week we're pulling and planting carrots, radishes, cilantro, and salad greens."

In front of Rosie are scattered photos of the farm's bounty in summer, showing tidy rows bursting with color and life. There's an aesthetic to the farm: neat, symmetrical, and not a weed in sight.

Rosie, tall and lean, is up and down, retrieving photos, pouring tea, and answering the phone while Chuck lounges on the couch, Mississippi drawl and salty language melting all pretense. The two will talk fondly of vegetables for hours, but their love for growing food doesn't eclipse their business sense: if an heirloom broccoli seed produces small heads, they're the first to scrap it for one that performs. "We grow what works in the big picture," Chuck explains. "This is a business. We could sell the *#%* out of green beans but the labor involved isn't worth it."

Chuck and Rosie don't have a 401(k) plan, but they make enough money in the summer to spend the winters sipping tea, telling stories, and reclining into couches, as well as enjoying their other diversions: powder skiing, rock climbing, playing thrash-country music, and writing about rural life.

Chuck and Rosie take a rare moment to relax at Stone Free Farm.

"We make a pretty comfortable living," Rosie says. They just bought an additional fifty-nine acres adjoining their property. They started with just three acres and a rototiller bought on loan.

On the sage plains in the arid shadows of the Rocky Mountains, temperature fluctuations cause tomatoes to sing in the daytime sun, then shiver by nightfall; rain is scarce and wind is plentiful; and the luscious mountain wilderness beckons all summer while the weeds grow. It takes quite a commitment, then, to be a successful farmer. It takes consistency, presentation, an ability to be good at many different things, and the discipline *always* to choose the weeds over the mountains.

Rosie and Chuck have made these choices at Stone Free Farm. They sell their harvest primarily at two local farmers markets: Rosie takes Cortez, while Chuck makes the trip to Durango. In nine years Chuck hasn't missed one Saturday market. By 7:30 he's set up, kicked back with coffee and newspaper, and taking the first breather since 4:30 that morning. That consistency of presence and product has led to a tradition. A line of people, money in hand, forms at both Stone Free market stands before the official opening at 8 AM.

Recognizing that people these days are used to picking their produce sparkling clean from supermarket aisles, Chuck and Rosie go the extra mile in presentation. Carrot tops are cut and the orange roots cleaned until they shine. Potatoes, radishes, beets, and onions are doused in water until all the dirt is removed. Greens and herbs go into a nonoperable washing machine where they're whisked in water, then bagged for sale.

The market stand is art, a still life with vegetables. Its colors, shapes, and textures reflect the artistry of their fields and vitality of their land. As eager customers line up, it is virtually impossible to imagine the relentless work that goes into such an offering. Summer workdays begin when the sun rises over the La Plata Mountains to the east. They are not over until it sinks behind the Abajo range in Utah, also visible from the farm. "We work six days a week, seventy to eighty hours a week," Rosie says. "We *make* ourselves take Sundays off." But even Sundays are not given to play in the mountains. "We stay inside with the blinds drawn and just sit. We don't even go to parties all summer," she laughs.

Rosie with a customer at the Cortez Farmers Market.

Every year the farm is visited by people who just moved to the area, bought some land, and want to start an organic farm. Chuck and Rosie are generous with information in the belief that more local, organic farms will keep the farmers markets alive and prosperous. But many of these newcomers are never seen again. "People have romantic delusions of farming. They

Chuck with a flat of seedlings.

have no idea of the work involved," Chuck says. "If we added up what we make per hour we'd moan and cry. Why does a lawyer make $200 an hour when the farmer feeding people is always poor? It's a crazy-assed system."

"But we're happy with our lifestyle," Rosie inserts. "We don't even want to get bigger. If we could continue like this forever, we'd be happy."

Despite the challenges of farming in the Southwest, Chuck and Rosie love it here. Chuck laughs about the fact that his family "can garden the hell out of an acre of land in the Southeast, without ever having to think about irrigation. But the bugs will eat you and your crops alive."

Although Chuck and Rosie sell more volume in the more affluent community of Durango, they appreciate the working-class, unpretentious nature of Cortez. "You might not want to talk religion or politics with some of the locals," Chuck warns, "but they'll go out of their way to help you every time."

The farm has always been organic. "We'd never think of using chemicals on the land we love," Rosie says. "I wouldn't eat food that was full of poisons and I wouldn't want anyone else to eat it either." The payoff for being organic has also been a balance in the ecosystem. Instead of chemicals, their defense is based on "herds" of preying mantises, scores of ladybugs, and armies of earthworms.

Rosie and Chuck believe that most anyone who is willing to work for it can have a successful farm here. It doesn't always take a lot of money to start up, but rather an unwavering commitment to stick with it until it begins to pay off. It also helps if you're good at bookkeeping, irrigation installation, marketing, soil science, greenhouse construction, and entomology.

Rosie says that she'd like to balance out the annual workload some, instead of overload in summer and standstill in winter.

"Maybe we'll get some cows," Chuck muses. "I don't mind cows too much."

THE STORY OF OUR INTENTIONS
Lost Cabin Ranch

Chino Valley, Arizona · By Susan Lamb

Rebecca Routson

"I really believe," Rebecca Routson says, "that every square foot of this earth requires its own particular attention."

The particular piece of earth Rebecca and her husband Don attend to is an organic farm in the Williamson Valley northwest of Prescott, Arizona. It is a place of unconventional beauty, set in low foothills on the edge of a dry and shrubby grassland dotted with junipers. A visitor to the farm is soon surrounded by a surprising riot of local vegetation that scarcely appears cultivated.

Rebecca is a lean, tanned woman, light on her feet and with a perpetual smile that could dazzle the sun. She thinks a lot about everything she does and is capable of giving up on years of hard work when she sees that there is a better way to do things. What she likes to talk about, in a way that indicates she's thought about it a lot, is what she calls "the story of my intentions."

"I have a master's degree in nutrition and taught at the University of Wyoming for a few years," she says. "The knowledge I gained through my academic career led me to realize that if I was going to eat right in this part of the country, I'd have to grow my own food."

Rebecca and Don moved to Arizona and spent a year looking for the right piece of land. After they found it, they talked to the farmer next door who advised them to raise squash and cucumbers – a crop they could harvest by hand.

In the early years, Rebecca and Don were too poor to own a tractor. They used shovels instead. They started small, with two acres, and added two more the next year, and two more after that. When their produce ripened, they picked five thousand squash in a morning and drove it to supermarkets in Phoenix. Their vegetables were organic, but there was no market for organic produce in those days. After a few years, they realized that they couldn't make any money selling squash at three or four dollars a box – "the box alone cost a dollar!" Rebecca exclaims.

Don got a job and Rebecca continued to experiment. She shifted from raising produce for the market to growing food for her family. Instead of growing squash and cukes, she cultivated things that would feed them all year – potatoes, onions, garlic, beans. Rebecca was a vegetarian when she was a student, but her health deteriorated when she started doing hard outdoor work. Don told her she needed a good steak; the

Rebecca Routson demonstrates her "chicken truck."

farm needed manure. They got a cow. She had that steak and felt a whole lot better.

Rebecca and Don began producing half vegetables and half grass-fed beef. Rebecca discovered that managing twenty acres of pasture is much easier than growing vegetables. The sale of their cattle began to pay for pumping irrigation water and for incidentals. The manure fertilized the crops.

Rebecca and Don have three children: Kanin, Cody, and Rafael. Rebecca notes, "Other kids are asked to take out the garbage, clean their rooms; there's not a whole lot of pride in that. When children are raised to be responsible for their own sustenance – feed the chickens, cut hay – there may be hay dust all over you, but you're feeding animals that will feed you. The circle is complete." As Rebecca puts it, "the farm is family dependent and the family is farm dependent."

It is this sense of completeness, of wholeness and interdependence, that constitutes the beauty of the Routsons' farm. They grow a minimum of row crops, preferring to plant in small patches scattered around low hills, in niches protected from the wind. Wild and cultivated plants support each other and blanket the soil to prevent evaporation. These microclimates can add two weeks to the growing season.

In the row-cropped bottomland, Rebecca sows clover in tractor-wide strips. The clover enriches the soil with nitrogen, prevents moisture loss, and diverts gophers and rabbits from the tomatoes and beans. For her fiftieth birthday Don made her a "chicken tractor" – a long wire enclosure on wheels that enables her to move chickens along these rows of "green manure." Their diet of greens, insects, and a little grain causes the chickens to lay eggs with day-glow golden yolks rich in Omega-3 fatty acids.

The Routsons' farm lies at the upper end of the large

Root cellar at Lost Cabin Ranch.

underground Chino Aquifer. They have no shortage of water. Rebecca uses a well-devised sprinkler system to water her plants. She explains, "Drip irrigation doesn't grow things like I want to grow them. It limits me to row crops and there's so much plastic garbage. Drip also confines the water to a narrow strip that concentrates the gophers on your crops." Rebecca continues, "If sprinklers are done properly, if you make sure that everywhere the water falls there's something useful growing, it's different. You just don't water on a warm, windy afternoon, but in the morning or at night."

Rebecca learned how to farm her land mostly by this sort of experimentation. Then she was asked to teach a class on agriculture at Prescott College, and she read some books to prepare for that. She realized that she could learn a lot from books, too. But she still holds with the old saying, "The best fertilizer is the farmer's footsteps."

"Farming will keep you in poverty," Rebecca says. "Every cent you earn is sucked up by the farm and it's very labor intensive. But the rewards are so great: the closeness of our family, the work ethic, the meaning to life."

A few hours of wandering in the Routsons' Eden are enough to make anyone a convert.

RETURN OF THE PEACHES
Ferrell Secakuku Restores Hopi Orchards

Sipaulovi, Arizona · By David Seibert

They may receive scant and unpredictable rainfall, but the Hopi mesas of northern Arizona support some of the oldest continuously inhabited communities in North America. When the Spaniards arrived in the region almost five centuries ago, the Hopi were agriculturalists. They recognized the peaches the Spaniards brought as excellent additions to their diet, and quickly learned to grow them both to eat and to use in their traditional ceremonies.

Hopi subsistence agriculture continued largely unchanged well into the twentieth century. However, increased contact with the dominant culture brought an influx of wage labor and packaged commodities after World War Two. Hopi agricultural practices began to decline. By 1951, many Hopi had discontinued farming altogether. Today, this shift away from dryland farming and toward a wage economy has changed the fabric of Hopi culture. The loss of their organic, local foods system has resulted in increased diabetes, decreased physical activity, and the disappearance of heirloom crops that support, and are supported by, pollinators and other insects vital to ecosystem health. For the Hopi, what we commonly call "ecosystems" includes humans as a fundamental element; the two remain inseparable both for physical health and for spiritual well-being.

Despite the current scarcity of dryland farming in Hopi land, peach trees can still be found. Their roots run deeply and widely through Hopi culture, and they represent much more than a supplement to the traditional diet. Named *sipala*, the fruit also symbolizes strength, endurance, and the ability to overcome hard times. Nowhere is this more true than in the village of Sipaulovi on Second Mesa. This is where Sipaulovi resident and Hopi elder Ferrell Secakuku has chosen to make his stand against the tides of change.

Ferrell Secakuku

Ferrell believes a return to ancient agricultural practices holds the key to the health of the Hopi people, their lands, and their traditions. As he points out, these three elements cannot be spoken of independently. "If I could revive the springs and the peach orchards and some of the farming that we had to go through to learn the meaning of the *Katsina*, to learn the meaning of these social dances, then perhaps Hopi youth could really know what Hopi is about," he says. (In Hopi stories, the Katsina are the spiritual beings who bring moisture and all good things to Hopi land when they return from the sacred San Francisco Peaks each year.) By bringing back agriculture, Ferrell believes he can share what he learned as a young person. He says, "At least the youth will have the experience, will see the experience of what I did when I was a kid their age, which [otherwise] they will probably never have. That's what I want to put a little bit more meaning behind, because after all we are Hopis, and they are supposed to know this."

Well into his sixties, yet sturdy and stronger than most people half his age, Ferrell continues to live, farm, and teach in the Hopi way – a way that remains committed to the health of all the world. Each time he visits the fields, he looks out over rolling hills once covered with fruit trees, squash, corn, beans, and sunflowers. Now he sees abandoned peach houses where tools and food were stored, nearly dead peach trees, and bare rocks where peaches once dried in the northern Arizona sun. Because of this neglect, many heirloom varieties once grown here are in danger of extinction. So is the traditional and sustainable way of living on the land that remains integral to the flourishing of both crops and the people who grow them.

Yet what Ferrell sees on the mesas today is the potential for reversing this trend. In the spring of 2003 – twenty days after the vernal equinox in accordance with Hopi cultural practices – he organized a large planting party involving community members aged five to seventy-two. The participants planted Hopi blue corn, squash, beans, sunflowers, plums, and of course, new peach trees. In addition to the familiar rewards of growing such crops, the group has an eye toward economic payoffs for the village by marketing some of the produce as

non-genetically modified, local, and traditional food. With the work begun, village residents will maintain the fields and crops and at the same time provide invaluable demonstration activities and workshops for future growers on the mesas.

The past and future merge in this old/new agriculture. Modern innovation, Ferrell feels, need not extinguish older ways. Small brush weirs and larger dead-wood gabions, or water catchments, in nearby washes slow water runoff during large floods, allowing nurturing moisture to reach the roots of the trees. Just as important, these simple woven structures hold back nutrients on their uphill side. Each gabion prevents topsoil from being washed downstream. The results are there on the ground to witness. "This is where we plant next," exclaims Ferrell as he points excitedly to the rich topsoil held in place at one silt-collection site. "This is where the good stuff is – and we don't lose it if we just pay attention to the flow." He likes the metaphor, both for work on the ground and in the community he calls home.

Ferrell hopes that his peach orchard project will serve as a model for other Hopi villages and for other peoples on the Colorado Plateau who want to protect their cultural and spiritual traditions. For him, restoring the culturally significant springs associated with orchards, conserving Hopi heirloom crop varieties, and encouraging a return to traditional agricultural practices all represent small steps of a larger journey toward community health. Many village residents would still prefer to get packaged foods rather than work in the fields. This, as Ferrell puts it, is simply an obstacle that the Hopi need to confront daily, rather than pretend it will go away or that it is inevitable.

Volunteers prepare the soil for planting a peach tree at the renewed Hopi orchard.

Giving up is not an option. It never has been, for the Hopi. The restoration of traditional agriculture involves, and from the looks of it requires, community outreach and information-sharing within and between Sipaulovi and other Hopi villages.

But nothing makes an impact quite like a peach tree loaded with bright pink blossoms in springtime. "This is it," says Ferrell. "This is what people need to see – that we can do it if we are willing."

CAPTURING THE BRILLIANCE OF PLACE
Sutcliffe Vineyards

Cortez, Colorado · By Roger Clark

John Sutcliffe

Tucked below the east face of Battle Rock in the far southwestern corner of Colorado, John Sutcliffe struggles for control of McElmo Canyon. Growing grapes at 5,300 feet above sea level isn't just a matter of turning sunlight into sugar, but rather also part of his strategy against what he calls a "new wave of settlers bent on remodeling the West into an endless playground."

More than a decade ago, John and his wife, Emily, purchased a small hay farm that came with water rights dating back to the 1880s. Their original intent was to use the farm as winter range for a small herd of cattle that would be moved to higher pastures on Bureau of Land Management property during the summer. Their plan was to fuel John's passion for working with livestock and the land, and meet Emily's desire to deliver babies for women who live in rural places.

However, the bureaucratic process required to maintain grazing permits on public lands soon tried John's patience. His solution was to combine his earlier experiences of farming in Wales with his knowledge of the fine foods industry that he had gained as a successful restaurateur. John became a grape grower and vintner. With characteristic determination and intensity, he now produces more than a thousand cases of fine wine each year.

John thrives on learning subtle and painstaking lessons from the land. "We haven't gotten to where we are from our wealth of knowledge and experience," he deadpans. "Of the 1,500 vines that we planted in 1995, we have fewer than 800 left." He rails out a litany of calamities, beginning with dishonest nursery owners in California.

At first the Sutcliffes concentrated their plantings along McElmo Creek. "I thought that because our cows went to the lowlands along the creek to deliver their calves that meant the microclimate in those areas might be milder," John says. "But I was dead wrong. The cold air draining down-canyon from Ute Mountain means that there is on average a nine-degree difference between the high end of my property and the part that is only 120 feet lower. That means that the last frost on the high end is about the 8th of April, and May 9th on the lower."

John's on-the-job training and hard work have been shared with Francisco Castillo, an immigrant farmer from northern Mexico. Together they maintain a grueling annual cycle. It starts in the spring, when they begin devoting at least twelve minutes to pruning each of the 8,000 vines that they now have. That is followed by all-night vigils of being prepared at a moment's notice to ignite smudge pots that provide enough heat to protect the plants from frost. The growing season's endless chores further include monitoring and maintaining plastic drip lines and periodically adjusting flow rates for each plant to assure optimal growth and sugar content. Enormous bird-proof nets are draped across the vineyards at the onset of fruiting. As harvest nears, John and Francisco thin leaves to allow just the right amount of sunlight on the grapes, and continually test the nutrient content of the leaves and the sugar content of the fruit.

Both growing grapes and making wine require close attention to detail.

A Sutcliffe Vineyards label on a bottle of 2002 Merlot reads: "Sun-drenched days and fine red soils make McElmo Canyon a wondrous garden of peaches, apricots, melons, and chiles. Deep in this canyon, between towering sandstone walls, lies Sutcliffe Vineyards. Perhaps we have captured the intensity and brilliance of this place in the color, depth, and complexity of our wine. Grown, produced, and bottled on the property."

John markets his wine only to a few local distributors and fine restaurants. When asked whether he's making a profit, he defers the question with "I'll tell you this, being married to a gynecologist is the key to my success." He shifts the conversation to the importance

John savors autumn when conditions of "jolly cold nights and warm days give the grapes the acids needed to ripen and have a liveliness in their taste, like an apple with a bit of a snap." Harvest is completed in a marathon few days. With the help of a dozen more workers, the fruit is picked and carted to stemming and crushing processors. The art and science of bringing out the best the grapes have to offer continues through what seems an endless series of steps before bottling. "There is a constant need for tasting, smelling, and adjusting," explains John. "You are up to your elbows in the process."

of keeping farming and ranching in McElmo Canyon. "This valley has been producing beautiful melons and vegetables for decades," he says. "The miners in the mountains above here would have starved without this place. But because people no longer value our products, the only bulkhead against the developers is an alliance between the farmers and our community. If we're not here, then the land will wind up producing nothing but mere playgrounds for the prosperous of our cities. Developers are just like coyotes during calving season, circling and waiting to seize the moment."

Careful husbandry, along with a salubrious microclimate, have helped Sutcliffe Vineyards produce superior wine grapes.

John's crusade for keeping the land in farming includes retiring development rights on his property. Sutcliffe Vineyards was recently commended by the American Vintners Association for "taking a leading role" in protecting the natural environment. In noting its donation of an agricultural easement, the *Durango Herald* cited Sutcliffe Vineyards for contributing to a "growing trend in locally based agriculture [that] provides for the continuation of agricultural and agribusiness uses of the land [while] preventing any future subdivision that would compromise the land's agricultural values or harm its resident wildlife."

In a guest commentary for the *Denver Post*, John wrote: "Just as stewardship is more than simply farming the land, farming is far more than a picturesque backcloth for sprawling homes. It contains the essence of the community and is a vital component in any community's ability to avoid urban sprawl."

An intimate sense of place accompanies John in his vineyard. Every row and every plant has a story that he passionately shares. And he proudly expresses the vitality of his community when he tells how Francisco recently paid him cash to purchase his own piece of the vineyard, and a home for his family.

HEALING TRADITION THROUGH INNOVATION

Diné, Inc.

Seba Dalkai, Arizona · By Roger Clark

The sun has barely risen as Morgan Yazzie walks out to his fallow cornfield in Chandler Springs Valley on the southwestern edge of the Navajo Nation. It sits in an arid landscape, treeless for miles, that to the untutored eye looks profoundly ill-suited to agriculture. Yet in this sun-baked place Morgan and other Navajo – or *Diné* – farmers are addressing modern problems through attention to both tribal traditions and modern innovations.

With Morgan is Hank Willie from nearby Seba Dalkai. After the traditional "*Yá'át'ééh*" greeting, the conversation turns toward the weather and the extended drought that has gripped the Southwest in recent years. On this day, the soil underfoot is still moist from a late winter storm, but snowfall on the shining San Francisco Peaks to the west is less than half of what's normal for the season. "Last year at this time," Hank notes, "things were looking pretty good. But then the spring winds dried out the soil and some people decided not to plant anything."

Morgan Yazzie has been growing food and helping neighbors do the same in this high desert country for nearly thirty years. In 1999, he joined others in creating new educational and

food-production opportunities through an organization called "Developing Innovations in Navajo Education, Inc." Diné, Inc. began its work by successfully obtaining funding to build and administer the construction of a new elementary school at Seba Dalkai. It broadened its scope by developing a community education program centered on nutrition and agricultural sustainability from a traditional Navajo perspective. Today the group's work includes demonstration farms, advisory councils, educational workshops, and agricultural curricula for local schools.

Hank Willie and Morgan Yazzie

Hank Willie serves as coordinator of the Diné Community Food Project that assists and supports Navajo families in the communities of Teesto, Dilkon, Birdsprings/Leupp, Tolani Lake, and Whitecone/Indian Wells in revitalizing dryland agriculture. The project's goals are to meet the food needs of

FARMING AND MARKET GARDENING

low-income families; to increase these communities' self-reliance through growing their own food; and to foster an appreciation for healthy foods and a healthy lifestyle.

Driven by drought, Diné, Inc. participants have adapted traditional practices when necessary. "Mother Nature and the elders are my mentors and teachers," says Hank. "We've been advocating drip irrigation as an alternative to traditional dryland farming only because of the continuing drought situation. We don't want to replace traditional agriculture, but to promote it." Even Morgan has added a drip irrigation system to water beans and other vegetables growing near his house on the hillside overlooking the valley.

But Morgan's unirrigated cornfield has remained productive despite the drought. The field is located in the valley bottom, where it catches runoff from occasional thundershowers that usually begin in July. He points to a section of the field where rainwater flowed in abundance during last year's growing season. "We had ears of corn that were over a foot long from there," he says. The multicolored corn he plants is from seed well adapted to the harsh growing conditions. It has been handed down from one generation to the next, and ultimately is descended from genetic sources that pre-date modern agriculture.

As is the tradition, he plants seeds in clusters spaced about a yard apart. The depth of planting varies with the depth of moisture in the sandy soil, where seeds will germinate and develop an extensive root system in advance of the late summer rains. He thins the clustered corn plants as needed. As the corn matures, it looks more like an orchard of bushes rather than the densely planted rows of sweet corn growing throughout the Midwest. When the corn ripens, family members and neighbors help harvest it by hand, keeping some of what they pick for food and ceremonies. If any is left over, it is shared and bartered.

Integrated with these traditional practices are more modern techniques. One of Diné, Inc.'s projects funded the purchase of a tractor for use throughout the community. Morgan drives and maintains the tractor and, in April, begins plowing fields for farmers in outlying areas. "Morgan didn't get around to planting his place until June last year," Hank notes. A raven lands on a nearby fencepost, and another arrives soon after. "They're scouts, always traveling in pairs," Morgan says. "I pull my horse trailer into the field when the corn and squash appear to make the ravens and coyotes think someone is living here." The field is fenced to keep Morgan's horses and cattle out until after the harvest; then they're penned inside it to feed on the corn stalks and to fertilize the field.

With the exception of a few head of livestock, food produced here is not sold to outside buyers. Rather, it is a tangible continuation of ancient traditions that rely on intimate local knowledge, as well as dedication and attention to the appropriate songs and prayers. As one Diné, Inc. report puts it: "traditional agriculture is not driven by economics, but rather a commitment to sustaining life, kinship, and tradition."

Left: The Diné, Inc. project combines traditional Navajo agricultural practices with modern techniques to help feed area residents.

The work of Diné, Inc. is only a small step toward solving serious economic, social, and health problems within the Navajo Nation, where some 60 percent of residents live under the poverty rate, the employment rate hovers around 50 percent, and median annual income is $4,500. Roads, utilities, telecommunications, and housing are in poor shape. Yet over the last century the Navajo people have become less self sufficient. Communities that once raised much of their own food no longer do so. This has resulted both in a departure from traditional diets and in a decrease in physical activity – which in turn have led to an increased prevalence of diabetes and of many other acute and chronic health problems.

The challenges seem enormous, but, as Morgan closes the gate and exits the field, so are the possibilities. Morgan needs to go to a healing ceremony for an ailing friend. Hank is headed to a health fair over at the Dilkon Chapter House. They speak briefly about their plans to develop gardens near homes for elderly Navajos "to give them something to do with their grandkids." It is not only heirloom corn, but also new hope, that springs from the sandy soil here when it is tended with care.

Morgan Yazzie

FARMING AND MARKET GARDENING

HEALTHY SOIL, HEALTHY PEOPLE
Ashokola Gardens

Snowflake, Arizona · By Patty Kohany and Charlie Laurel

You have to slip into four-wheel-drive to negotiate the descent from the mesa top to Ashokola Gardens and Peacefield, the remote garden, homestead, and sustainable-living center where Kim Howell-Costion lives with her husband, Joseph Costion, in the high piñon-juniper country east of Snowflake, Arizona. It's worth every bump to get a glimpse of a place where this pioneering couple has honed the crafts of sustainability to a high art, one that inspires not only them, but numerous neighbors too.

Kim came to this place seventeen years ago and began caretaking a one-room cabin without plumbing or electricity. The path down off the mesa was impassable for cars. Everything, at times even water, had to be carried down in five-gallon buckets. Digging by hand, she planted garden beds irrigated with rainwater she had collected off her roof. In the early days a pick would practically bounce off the hard clay. Kim added compost – lots of compost.

Determined and independent, Kim hitchhiked into Snowflake for supplies now and again. Kim's sister introduced her to Joe. He was taken with this maker of soil, master gardener, wonderful cook. "She was committed to the land," Joe says now. "She had a dream and was gonna make it work." When he proposed to her, he promised he'd never ask her to leave the land.

Joe is the director of the Industrial Technologies Program at Coconino Community College in Flagstaff. He served on a nuclear submarine during the oil crisis of the early 1970s, an experience that fostered his undying dedication to the cause of renewable energy. Joe wanted to build an energy-efficient home in order to put his principles into practice. "Buildings use 50 percent of our energy supplies," he points out. "We could have an impact on 50 percent of our energy use by teaching energy efficiency in building construction."

Kim Howell-Costion and Joseph Costion

Kim, wanting to keep things simple, was resistant at first. Joe won her over by offering to build a root cellar. The two-story home is built in southwestern style, with wide porches and exposed vigas. Joe built the place using passive solar design principles. He used concrete blocks, grouted solid and covered on the outside with rigid insulation, for the exterior walls. The outer insulation layer allows the inner thermal mass of the concrete walls to store up the sun's warmth in the winter and coolness in summer. Roof overhangs shade out summer

Through constant experimentation, the Costions have made a wildly successful, and sustainable, garden and home.

sun. Natural light from carefully placed windows reduces the need for artificial lighting. Solar panels power lights and a few appliances; solar heaters warm water for washing and bathing. A "Watson Wick" graywater system recycles precious water.

The south side of the house is an 800-square-foot solar greenhouse that serves both to heat the house in winter and to get Kim's seedlings started. She starts some thirty thousand seedlings a year. Half are sold through a Community Supported Agriculture (CSA) program in which subscribers sign up to receive a shipment of seedlings once a month in spring and summer. She plants the other half in her own one-acre garden. The greenhouse pays for itself in CSA income, and in reduced heating costs for the household.

After years of intensive work, the soil at Ashokola Gardens is rich and lush. "These high desert soils can seem really challenging, but you really can grow bounty out of them, and they continue for a long time," Kim says. She has experimented with various ways of building soil, but often uses the so-called lasagna method. She plants a green manure crop, then turns the crop into the soil and puts a layer of waterlogged corrugated cardboard on it. Then she adds a thick layer of leaves, a thinner layer of manure, and a thin layer of pine needles. A bed layered in this manner in the fall is ready for planting in spring. To plant, Kim digs a hole in the cardboard and plants seedlings in the soil.

Kim estimates that she composts twenty to thirty tons of organic material each year. She is on the lookout for appropriate material such as leaves and pine needles all the time. She puts it in piles eight feet wide, some twenty feet long, and four to five feet high. Its breakdown is fueled by appropriate amounts of water and by copious numbers of worms; Kim regularly checks the temperature to make sure each pile is right for the worms.

Along with making good soil, Kim manages microclimates in her garden. She plants corn, squash, and beans in clusters in which plants shield one another from sun and wind. She shades rhubarb from the intense sun with grape vines that thrive in the new, deep soil. Fences between beds stop both animals and wind and support berry crops. Leaky dams built of leaves, pine needles, twigs, and branches slow the flow of water in washes, stopping erosion and allowing sediments and nutrients to enrich the soil. She and Joe have built retaining walls that soak up solar radiation and give it off slowly at night, buffering plants from the day-night temperature extremes typical of their arid climate and 5,600-foot elevation.

The house, outbuildings, and gardens demand continuous work, but Kim and Joe also devote a great deal of energy to teaching. Joe teaches courses in passive solar design, innovative building methods, solar greenhouse construction, and other energy-efficiency alternatives. He teaches his students how to calculate heat transfer and insulation and thermal mass values, but he also teaches the importance of aesthetics: the beauty and comfort of good buildings. For Joe, efficiency, elegance, and concern for what ordinary people can afford go hand-in-hand.

Kim, for her part, teaches a series of gardening classes in Snowflake and Flagstaff and classes specific to growing and cooking with such crops as onions, garlic, chilis, tomatoes, and basil. She's helped hundreds of people learn how to cope with the often difficult conditions of the high desert. "Healthy soil," she says, "leads to healthy plants, which leads to healthy people."

If Joe the builder and Kim the gardener and cook are any indication, it's certainly possible to craft a healthy life in the high desert very much in keeping with the principles of simplicity and stewardship for the land.

WATER AS LIFE
The Masayesva Family

Black Mesa, Arizona · By Gary Paul Nabhan

Victor Masayesva, Sr.

Patuwaqatsi. When Hopi elders say "water is life" it is not a cliché but a fact of life.

Anyone who doubts that the Hopi truly honor water as something that tangibly sustains their lives should visit a farm tucked back in the headwaters of Oraibi Wash, between two fingers of the outspread hand formed by Black Mesa in northeast Arizona. There Victor Masayesva, Sr. still cultivates several rainfed acres as he has done for decades. His blue and white flour corns, sweet corn, beans, squash, watermelons, cassava melons, and fruit trees need constant tending.

Victor still walks out into his fields at dawn, tending to his crops, weeding out their competitors, and chasing away ravens, regardless of the droughts, plagues, political changes, and economic pressures that have forced many American farmers to go belly-up. In fact, the only change in his routine of the last half-century has been a rather recent one. After his fiftieth wedding anniversary with his wife, Victor decided to give up sheepherding in order to devote himself full-time to farming. In the midst of the worst drought in more than a millennium, the fields of neighboring Navajos, Anglos, and Hispanics are stunted and suffering, but Victor's dozen acres of rainfed fields of blue corn are tasseling out and maturing full ears. His other crops look just as lush.

It is not merely his sustainable cultivation of crops that makes Victor's example so inspiring, but the follow-up work that he and his wife do to prepare traditional Hopi foods. He has constructed a huge stone-lined pit in which his family seasonally roasts a half-ton or so of sweet corn. He has built a *piki* house in which he stores seeds and his wife kneels down weekly to make wafer-thin blue piki bread in the traditional manner. His crop seeds are saved from year to year, and backed up in long-term storage at the Native Seeds/SEARCH germplasm facility in Tucson. They are seeds specially adapted to the Colorado Plateau. His blue flour corn has seedlings with extra-deep roots to reach into soil moisture and extra-long hypocotyls that can emerge from planting as deep as twelve inches below the ground surface. His mottled lima beans have root knot nematode resistance. His tepary beans show extraordinary tolerance to drought, heat, salinity, and alkalinity. In short, they are exquisitely adapted to the prevailing conditions of the plateau and able to survive on as little irrigation as any crops in the world.

Water is a big political as well as cultural and agricultural issue for the Masayesva family. Every one of Victor's children has forged a career that has somehow touched upon water and its role in native language, culture, subsistence, and survival. Two of Victor's most successful Hopi crops are his sons Virgil, founder of the Institute for Tribal Environmental Professionals, and Vernon, founder of Black Mesa Trust. The latter is founded on the idea that, as its educational materials state, "Water is not a commodity to be bought, sold or wasted.... Water is sacred, especially in the Black Mesa region where water is key to our survival."

Vernon Masayesva and other Hopi leaders make these pronouncements as practiced natural resource managers who know how to make the best of the meager moisture hidden in pockets within the stretch of the Painted Desert they call home. Increasingly, however, they repeat these ancient aphorisms cognizant that their land is now drier than it has been within the collective memory of their cultural community. Climatologists agree. At the beginning of the twenty-first century it appears that the canyon country of the Colorado Plateau is suffering from the most prolonged, severe drought in 1,400 years. Not only has rainfall been unusually spotty, but winter snows have

Over generations, indigenous farmers have produced myriad varieties of corn on the Colorado Plateau.

melted quickly, wildfires have ravaged upland watersheds, and most freshwater springs have all but dried up.

These indicators of drought are not interpreted by the Hopi merely as physical changes in the landscape, but also as signs of an imbalance between humankind and the natural world. Most Hopi recognize that, even during the worst of other periods of limited rainfall, a trickle of water still dripped from their springs. Today they ebb because of the accelerated pumping of the Navajo Aquifer – the sole source of drinking water to the villages on the Hopi Reservation and to many ranches on the Navajo Reservation. It has also been drawn upon for the past thirty-five years by the Peabody Coal Company, which has pulled as much as 1.3 billion gallons out of the ground annually in order to slurry coal from its Black Mesa mines in a 273-mile pipeline to the Mohave Generating Station in Laughlin, Nevada.

Peabody officially claims that its groundwater pumping has little to do with springs drying up when compared to the impacts of domestic uses and the drought itself, but many scientists and Native American elders think otherwise. Scientists cite the aggravated effects on well and spring drawdown, as well as ground subsidence in the area of Peabody's wells compared to other drought-stricken sites nearby. And Hopi elders see long-term consequences. As they explain in a statement regarding the groundwater pumping associated with Peabody's mines, "Water under the ground has much to do with the rain clouds. Everything depends upon the proper balance being maintained…. Drawing huge amounts of water from beneath Black Mesa in connection with strip mining will destroy the harmony…. Should this happen,

our lands will shake like a Hopi rattle: land will sink, land will dry up, plants will not grow, our corn will not yield and animals will die…."

Victor's son Vernon has further explained the critical significance of springs in Hopi cosmology: "Springs are the breathing holes between the Fourth World that we currently live in, and the earlier worlds we emerged from. What happens to springs affects our future as a people." Unfortunately, Vernon and Victor know all too well what is happening to these breathing holes. Over the last half-century, they have witnessed the drying up of more than three-quarters of all the springs within walking distance of their farm along Oraibi Wash. The water depth in wells has decreased by a hundred feet or more. The rate of their loss, Vernon contends, has accelerated since Peabody began pumping from the Navajo Aquifer around 1970. He refers to this trend with the Hopi term *paatski*, "the tearing up of water."

Peabody's pumping of groundwater from the Navajo Aquifer may soon be curtailed. If that happens, Vernon will look to his father's example as a guide to what he and other men should concentrate on next. The Masayesvas imagine what they call "a learning plaza" for sustainable use of natural resources – an outdoor school in which traditional Hopi teach their youth and their neighbors their traditions of adaptation to this dry, wind-blown land. If they are ultimately successful, the Masayesvas will have produced the ultimate blessing: a crop of younger people who will know how to take care of the land with the same diligence and acuity of vision that their elders have shown.

MAKING USE OF
NATURAL CONNECTIONS
Oakhaven Permaculture Center

Hesperus, Colorado · By Rachel Turiel Hinds

In another time and place Tom Riesing crunched numbers on Wall Street. Christie Berven taught elementary school. Since meeting in 1998, the two have become born-again zealots for their cause: soil, earthworms, beet greens. Tom and Christie are the creators of Oakhaven Permaculture Center, tucked into the Gambel oaks and lichen-covered rocks at 8,700 feet at the mouth of La Plata Canyon near Durango, Colorado. It consists of a 2,200-square-foot greenhouse, outdoor gardens, ponds, chickens, and the ever-watchful gazes of its creators.

Christie is high-energy exuberance and fire. She pins you with her eyes, talking so fast you hope she remembers to breathe. Tom is stone quarried from a deep, still place in the earth. His movements are slow and calculated, as are the thoughts he expresses. At the ages of fifty-seven and sixty-six, respectively, Christie and Tom are starting a new sort of family, and certainly a new sort of life.

Tom explains in his practical way that permaculture is an agricultural movement. It's a term coined by Bill Mollison of Tasmania based on the phrase "permanent agriculture," and has come to encompass the idea of a sustainable culture and economy through working with nature. Christie holds up a banner she made that bears the heading *Permaculture*. A circle on the inside holds the word *ethics*. Radiating out from it are four phrases: Care of the earth. Care of all beings. Share the surplus. Aware of the limitations of the earth.

Christie asks, "Have you ever seen a mountain meadow?" She points north to the La Plata Mountains, which cradle many such meadows. "There is a synergistic relationship happening. Some plants are taller than others, and those that need protection from the sun will grow near the taller plants. No one rototills, no one fertilizes; the leaves die in the fall and cover the ground, protecting it from the sun and adding nutrients. We study these natural systems so we can care for and benefit from the earth with similar ease and efficiency."

The permaculturist believes in working with the natural features of the land to produce more with less work. If you've

*Christie Berven and
Tom Riesing*

FARMING AND MARKET GARDENING

got a cold spot in your house, create a root cellar. If you've got a slope, grow moisture-loving plants at the bottom where rainwater will collect. If you've got oak trees where you want a garden, trim their limbs and use them as trellises to grow grapes, hops, and fruitful vines.

Put into practice, permaculture harnesses and recycles free water and energy. In this spirit, Tom and Christie collect rainwater and snow on the north side of their seventy-two-foot-long greenhouse, channeling it inside where it warms up in a large pond and then is used to water their plants. Heat, too, is sacred, and every bit possible is collected, stored, and re-released. Any tilling of the soil is done by earthworms, chickens, ants, and snakes, all of which are welcome in the greenhouse and outdoor garden spaces. To prepare a new bed in fall, Tom and Christie lay down cardboard, "which the earthworms *love to eat*," Christie says. That is topped off with six inches of manure and a thick layer of straw. When the rain and snow come, they help decompose the "compost sandwich" and keep the worms hydrated. By spring, all the work is done and they're left with a foot of excellent planting medium.

Tom decries farming practices in California, where great acres of strawberries are grown. "First they spray the land with methyl bromide, killing everything. Then they add fertilizer to support life." "It's all backwards!" Christie exclaims. "They're killing the insects, fungi, and microscopic organisms that naturally build and fertilize the soil. Plus it's too much work."

Left: Oakhaven's greenhouse works with the flow of nature by carefully recycling water and heat.

A basic principle of permaculture is to produce more than what you put in and to have each element of the design perform at least three functions. Tom and Christie's greenhouse pond illustrates this ideal. Its mass of water and concrete absorb heat during the day and radiate it at night. The pond holds and warms water for the plants, and increases humidity, which the plants love. "There's so much moisture in here you need an umbrella," Christie says. The algae that grows in the pond is scooped out and given to the plants as fertilizer.

In the middle of winter the greenhouse is like a temperate coastal farm. Fragrant *Nicotiana* flowers grow more than head-high, plump figs droop toward the ground, and a thigh-high mound of calendula seems to wave hello with its yellow and orange, sand dollar-sized blossoms. "Fairies live here," Christie announces matter-of-factly as Tom plucks dill, arugula, and beet greens to sample. Tom, Christie, and a motley, generous crew of friends and students from nearby Fort Lewis College built the greenhouse. It is heated at night, though only to forty degrees. Despite the chilly evening temperatures, tomatoes and chili peppers are steadily turning from green to red. Christie points out how the tomatoes grow in winter: low to the ground to conserve heat. "The plants are brilliant!" she observes.

Tom and Christie have big plans. They want to build a center for classes and workshops on permaculture design and sustainable building, eating, and living. It is a labor of love, says the ever-expressive Christie, grinning and grabbing Tom by the arm. "We're a perma-couple, Tommy and me, and all we need is lovage."

FARMING TO SUSTAIN COMMUNITY
Whipstone Farm

Paulden, Arizona · By Tim Swinehart

Cory Rade

Having cleaned chimneys for much of his life, Cory Rade decided about ten years ago to mix things up a bit and try farming. He was unfazed by the fact that he and his family had very little farming experience. Today, as he talks about how he came to love being a farmer, Cory retains some of the archetypal chimney sweep's good nature – a sparkle in his eye, excitement in his voice – as he describes how they learned to handle the soil on Whipstone Farm, north of Chino Valley, Arizona, and how he lives and works there today with his partner, Shanti Leinow.

"Everything we've learned has basically been from books and from experiments," he says. "It's one grand experiment, really; every year that's what we do." It took about seven years just to "get the soil right," which has involved a long process of soil enrichment with manure inputs, as well as many unsuccessful experiments with different irrigation systems. "Since we didn't have anybody in the area to teach us, we spent $1,500 to $2,000 just experimenting on

different types of irrigation systems." The result was a system of drip hoses that can be tailored to suit the needs of individual rows of crops, including peas, beans, lettuce, spinach, radishes, turnips, beets, herbs, cucumbers, squash, eggplant, tomatoes, onions, garlic, shallots, and more.

The crops at Whipstone are grown organically. That's been a challenge too. "Our goal was always to go organic, but we didn't realize how hard it is," Cory says. "We didn't realize how many bugs there are or how many weeds there are. Trying to find ways to combat those has been one of our issues." Some of the bugs are eaten by the ladybugs and praying mantids he releases; others are distracted by "catch crops," such as potatoes, planted upwind of the most desirable crops. That keeps pests, such as the blister bugs that have virtually destroyed the alfalfa-growing industry in the area, away from the higher-value plants. When Cory discovered that field bindweed, one of the most common weeds in the area, has a high protein content, he began feeding it to animals on the farm, figuring that it might as well be used for something.

Right: Shanti Leinow and Cory Rade with a harvest of leeks from Whipstone Farm.

FARMING TO SUSTAIN COMMUNITY

Challenge and experimentation have also defined Whipstone's marketing. Not everything that grows well in the Chino Valley area is a broadly popular food, which is a challenge for farmers who sell directly to consumers through a community-supported agriculture project and at regional farmers markets. Recently, Cory got into fava beans in a big way. "We had to experiment ourselves a bit in order to tell people how to eat them and how to prepare them. We were lucky because the chefs at the market used them in cooking demonstrations, and that got people interested." Providing free tasting samples at markets, too, entices customers.

Squash blossoms at market.

Cory is president of the Prescott Farmers Market, and he has made extending its circle of community a priority. Sellers there have begun accepting food stamps and vouchers from the federal Women with Infants and Children program, both of which allow low-income shoppers to buy at the market. "People are coming down and they're experimenting with food," he says. "They'll look at stuff sometimes and say 'What is this? I've never seen this before... okay, I'll take some!' because they get twenty dollars that they can use to try something new. They wouldn't go to the grocery store and spend twenty dollars on anise or rhubarb or favas, but they come to the farmers market and somebody is there to talk to them about it. They make that connection and then try it out.

Cory likes selling directly to those who will eat his produce. "When people buy the food that I grow *from me* at the market, I have the feeling that they think of me when they eat it," he says. "Say someone eats some really good spinach and then comes back to tell me about it or tell a friend about it – that's a good feeling. At the grocery store you go get what you want and you bring it home, but you don't know how much work it took to get it there, what country it comes from, or how it was grown. There's no relationship to that food, it's just fodder. As my kids were growing up, we found that most of their friends' parents rarely cooked a meal. If it didn't come frozen or in a can or in a box, they didn't eat it."

"The customers get a community feeling just by supporting local growers. I'm sure they're thinking community when they're buying something. The thing that's amazing is that you've got these people that come out year after year. You get sixty-year-old people, forty-year-old people, and twenty-year-old people, and they're out there conversing and shaking hands. It's real community and real friendship, and everyone looks forward to that interaction each week." It's clear from the sparkle in Cory's eye and the excitement in his voice that he does, too.

GROWING MORE THAN VEGETABLES
Bob Kauer's Shared Harvest Community Garden

Durango, Colorado · By Charles E. Jones and Rose Houk

When Bob Kauer purchased thirty-six acres of an historic farm east of Durango, Colorado, in 2001, he wanted to live a dream. He would have Gaited Morgan horses frolicking in pastures (which he got), and he would raise a bountiful organic garden beside the Florida River. Even more, he desired a place where the community could tend and harvest copious amounts of produce, with enough left over to give away to charities. Thus, Bob provided an acre of land and the water for a garden where "we share the work and we share the produce."

Community gardens aren't a new concept, but in the Shared Harvest Community Garden the word "community" definitely deserves a capital C. Member households work the soil, plant the seeds, pull the weeds, train the beans, and perform other necessary tasks. An old dairy in Bob's barn has been converted to an office, and two managers visit once a week, keeping a work log of tasks that need doing. A small steering committee of garden members makes the seasonal decisions about what to plant and what techniques to try. "We're completely open to somebody who wants to try something," Bob says.

Each household labors in the garden, takes home as much organic produce as it needs, and shares modest expenses. After only two years of operation, with some forty households already participating, the steering committee determined that

Bob Kauer

the single acre of ground could support nearly sixty households. It's first-come, first-served as far as who gets to participate, and Bob happily reports that so far no one has come to harvest without sharing in the work.

Even with that many households carting away bushels of vegetables, the word "community" holds a broader meaning here. "Our goal," Bob explains, "is to always produce more than we need so that we have enough to give away to the soup kitchen, the women's safe house, and the homeless shelter. As long as we have excess to give away, that makes people feel good. People are drawn to the garden for that charitable purpose." To Bob's surprise, the garden has yielded enough food not only to feed many households, but also to deliver large quantities to worthwhile organizations. The one thing they do not do, as a policy, is sell the produce.

What motivated Bob Kauer to organize this garden? A Nebraskan by birth, he says he has always been interested in

Bob works with local gardeners who share the harvest with worthwhile organizations in the Durango community.

gardening and always had a small organic garden. Still, he wanted a big garden, but never wanted to do all the work and really didn't need the money. "This is a way that I can have that same thing: a great big garden and a bunch of people to get it accomplished, and I get to look at it every day, to live with it, and of course eat out of it too." Besides, he adds, "this kind of people are wonderful people to be around."

Upon arrival, everyone is greeted by Satchel, Bob's brown-and-white, fetch-obsessed Springer spaniel. The fence that encloses the acre surrounds many different textures and vibrant colors. Freshly weeded paths stripe the garden, while the other rows, 250 feet long, will be weeded before the end of the week. Green seedlings poke though holes in black plastic mulch that reduces evaporation and keeps down weeds.

Each year, several thousand seedlings are raised in a professional greenhouse by one garden member. Bob ticks off a litany of some of the crops that have been grown: seventeen different kinds of winter squash, six kinds of summer squash; six eggplant, seven pepper, and eight tomato varieties; many types of greens, "a lot of garlic," and even watermelon and cantaloupes – a real coup at an elevation of 6,500 feet in southwest Colorado.

In addition to using plant starts, the garden members installed an efficient drip irrigation system fed by Bob's well. Planting begins as soon as the ground can be worked, usually in April, with cold-weather greens such as kale, spinach, chard, and snow peas. The thin plastic mulch helps hasten ripening of warm-season crops like corn, squash, peppers, and eggplants. The garden continues to produce into October, and after a big harvest celebration it is "put to bed" in November.

A large old apple tree, enduring the years, occupies the center of the garden. Children's toys sit underneath, waiting for the next youngster who'll play while mom or dad works. Some children may help pull weeds, or explore the old hay barn, look for wildlife along the river, or relish the mysteries of the surrounding forest. Gardeners frequently gather for work parties or potlucks, talking, sharing food, and playing. They are a diverse group, from teenagers to people in their seventies. "We have people with all different levels of skills in the garden," Bob observes. "We have people that know very little about gardening, and then we have people with PhDs who teach horticulture at the college and others who are master gardeners."

Even the compost pile is a joint endeavor. Each household brings biodegradable wastes to build soil. Garden members collected several hundred bags of leaves from town to add to the compost. Local businesses have also provided materials. The garden is on former pasture land and so the soil was good to begin with. "We were very fortunate with our plot of ground," Bob explains, "because it had very few rocks in it. It is sandy loam with just a little bit of clay in it. It needed no amendments and had a pH of 7. But it had every weed seed that had been around here in a hundred years. It seemed that every weed seed germinated the first year when we started adding water, and we were almost overtaken by weeds."

Still, he was lucky, Bob says. The acre was already fenced, the soil was perfect, and there was a producing well right next to it. Bob put up the capital for the irrigation system, to be paid back slowly. Such an up-front infusion of money would likely be necessary if someone wanted to start a similar garden elsewhere, he observes – as well as "somebody who's really interested in seeing it happen."

Bob Kauer had the land, the water, the vision, and a philosophy of sharing. But after his initiation, the garden members from Durango households took over and made it happen. They've also created an even more seamless experience for local groups and school classes. "The young kids have come out to harvest and take the food to the homeless shelter, where they cooked it for the homeless," Bob adds.

Bob Kauer's ambitious, generous vision of building a beautiful community garden has quickly been accomplished. And it can serve as model for others who really want to put the word "community" into their own gardens.

SECTION TWO
Ranching

In arid lands, people have long survived by having animals do the eating of otherwise unpalatable plants for them. Raising livestock on rangelands has a long history in the Southwest, dating back to the early days of Spanish exploration – or even farther back, if you count the turkeys kept by ancient Pueblo dwellers. Ranching has been practiced by virtually all the groups of people living in the region – Native American, Hispanic, Basque, Anglo-American. At times it has dramatically harmed the land; in recent years it has often been an issue of contention in debates about the management of public lands.

Yet a growing number of ranchers are today practicing techniques that sustain both livelihoods and the land. Common threads weave through their stories featured here. All tend livestock, whether cattle, sheep or goats. Most run family-based enterprises. All are trying to maintain traditional practices, with an eye to a sustainable future. And in twenty-first century America, all face similar challenges: complex land ownership, grazing rights, development pressures, and uncertain markets ruled mainly by large-scale, international economics.

For an illustration, look to Billy Cordasco, president of the board of Babbitt Ranches in northern Arizona. He has under his purview close to 700,000 acres, land that has been in the family for more than a century. Every time board members must make a decision, Billy asks them to consider three things: economics, the environment, and the community. Two other Arizona ranch families, the Metzgers and Prossers, formed the Diablo Trust in 1993 to ensure that their operations stay economically viable and in harmony with the environment.

At a smaller scale, the Manzanares, of northern New Mexico, shepherd 1,000 sheep, including Churros, the oldest continuously produced breed in North America. The Manzanares family still moves these sheep from winter to summer pasture, and that herding system, they believe, is key to remaining in control of what they produce – certified organic and grass-fed lamb.

Churro sheep are a breed known to the Navajo, who have their own long tradition of sheepherding. For them, sheep traditionally have represented wealth, and they have been an integral part of their culture. Sheep Is Life (*Diné Be'iina*) and Black Mesa Weavers for Life and Land are working to revitalize the Churro breed. They are also teaching a new generation everything else associated with sheep – stories, songs, and good grazing practices. That's the sort of big-picture thinking that might just allow numerous small-scale projects in sustainable ranching to make a big difference.

SHEEP IS LIFE

Diné Be'iina and the Black Mesa Weavers for Life and Land

Navajo Nation, Arizona · By Gary Paul Nabhan

A Navajo weaver continues the ancient tradition.

When one first sees a flock of Navajo Churro sheep moving across the sage-covered flats of Navajo Nation lands, it is easy to imagine that they have been here, adapting to this land, since time immemorial. Their colors – buffs, browns, silvery-blues, cream, and black – seem to reflect the sky and the geological strata on the cliffs above them. They are the first and oldest continuously produced breed of sheep in North America. The ones on the Colorado Plateau today are probably descendants of those brought into northern New Mexico by the Oñate entrada in 1598, after their ancestors had adapted for millennia to the arid conditions in Spain, northern Africa, and the Middle East.

The progeny of the original herd of Churros brought up from New Mexico survived the Pueblo Revolt of 1680, after which many Puebloan families moved into Navajo communities on the plateau; the Puebloans, who had learned Hispanic weaving practices, then passed those practices along to the Navajo. By 1700, both sheepherding and weaving had already been widely adopted by Native Americans from the Rio Grande to the Little Colorado.

It is no wonder, then, that some Navajo believe that sheep have always been part of their culture, and assumed that their own flocks were somehow derived from the desert bighorn that persist in the arid canyons around them. Some Navajo recount that wool was a gift from the Holy People. In one version of the Navajo Creation story, weaving was first learned by Spider Man, who then taught Spider Woman, who then taught Changing Woman and the rest of the Navajo. As medicine man James Peshlakai was admonished by his grandmother to remember, "The blood running through the sheep that you're herding is the same that ran through the veins of your great-grandfather's sheep. Don't ever forget their energy. They will feed you, they will clothe you, and your sheep corral will be your bank account."

Those are the values that two small but courageous groups are reinstilling in Navajo youth in dozens of communities across the largest reservation in the United States. One of the groups

Right: Churro sheep are superbly adapted to the Southwest's arid landscapes.

SHEEP IS LIFE

is called *Diné Be'iina* (Navajo for "sheep is life"), founded in 1991 by the Begay family and their friends in the Ganado area. Diné Be'iina (DBI) has been widely successful in attracting an allegiance to its Sheep Is Life festivals for many years running. The other group, called Black Mesa Weavers for Life and Land, was founded in 1998 and is a project of the international organization Cultural Survival, Inc.

Working with Utah State University's Navajo Sheep Project and with a larger coalition of Navajo, Hispanic, Native American, and Anglo sheep producers called the Navajo-Churro Sheep Association, DBI and the Black Mesa Weavers have helped revive and improve the quality of this endangered livestock breed over the last decade. The breed had dwindled to a few hundred pure-bred individuals before Dr. Lyle McNeal of Utah State began a breeding program to restore the Churros to the vitality they had prior to the federal livestock reduction program of the 1930s. At that time, hundreds of Navajo sheep flocks were destroyed in an attempt to reverse desertification.

Churro wool bagged and ready for sale.

Fortunately, the Churros had enough unique qualities – long staple wool; tender meat and rich milk; and resistance to helminthic parasites, liver flukes, and ovine foot rot – that both the Navajo and their Hispanic neighbors in the Four Corners saw reason to work with McNeal. By some accounts, the Churros have now rebounded to a population of 3,000 to 5,000; many, however, are no longer the original purebred strain but crosses between Churros and Merinos or Rambouillets. The American Livestock Breeds Conservancy continues to list the Navajo-Churro as a Conservation Priority Livestock Breed.

But the genetic restoration of this sheep breed is only part of the story. Working with Black Mesa Weavers for Life and Land and the Institute for Integrated Rural Development at Diné College, DBI has fostered a revival of orally transmitted sheep cultural lore and traditional practices. The Navajo are relearning traditional songs about sheep, ways of building wooden corrals and lambing pens, and uses of traditional plant dyes to color their wool. As former DBI project director Malcolm Benally explains, Navajo rugs are not merely decorative; they embody many teachings critical to the cultural survival of his people: "There is a story to a Navajo rug," he says. "It's not a game. There are teachings in the weavings."

Colleen Biakeddy, field coordinator of Black Mesa Weavers, agrees with Malcolm wholeheartedly that there is something unique about the relationship among the Navajo, their original breed of sheep, and their weaving traditions. "We need to get this wool back into the hands of more Navajo weavers," she says, "for the traditional part it has always played in our lives. It was our dress. It was our ceremonial sashes. The knowledge is still there on how to make these things, but we're teetering on the brink of losing this knowledge."

The project director for DBI, world-renowned weaver Roy Kady, feels that this knowledge will persist among his people only if the younger generation is enriched with it from the very start. He takes DBI's "Spinoff" programs into elementary schools and even into Head Start programs, and has taught children as young as ten months of age to sit before a loom and weave.

For their parts, Colleen and Malcolm, both from the Hardrock Chapter on Black Mesa in northeastern Arizona, know that sustaining this cultural tradition also means moving back to rotational grazing practices to sustain the land itself. Colleen's vision is that healthy land fosters healthy communities, and vice versa. "We look for ways to keep Churro sheep nutritionally sound and healthy, and one way to do that is rotating their grazing through shrubs, grasses, and herbs," she says. "Navajos did rotational grazing a long time ago, even granting permission to let others graze across their own traditional territories to help their neighbors. Then, fifteen to twenty years ago, many herders walked away from that tradition, so now we have to bring it back."

Malcolm's personal vision of how to stimulate the return to sustainable grazing practices is by embedding the "Sheep Is Life" philosophy into a new initiative on Navajo Nation lands. The initiative, set in place by the tribal council's passage of the Local Governance Act, calls for each chapter house to devise a land use plan for common lands in its area.

"We need to make sure that our cultural teachings about sheep inform those plans," Malcolm suggests. In his view, the goal is not only to learn the land stewardship practices of elderly sheepherders who were excellent stewards of the land, but also to learn from the sheep. "I've learned a lot from the sheep themselves," Malcolm says, chuckling. "Every time I think I've learned everything about sheep, they teach me that there's more to it."

DOING WHAT HAS TO BE DONE
Antonio and Molly Manzanares

Los Ojos, New Mexico · By Sue and Tony Norris

Molly and Antonio Manzanares

Situated in the high country of northern New Mexico, in the lush Chama Valley a few miles outside the quiet village of Los Ojos, is the ranch of Antonio and Molly Manzanares. They are the owners and operators of Shepherd's Lamb, which produces certified organic, grass-fed meat and wool. With a flock of more than a thousand head of Navajo Churro and Rambouillet sheep and four children, you'd expect the couple to be busy, but that's just part of the story. Seeking to maintain an agricultural base in their ancestral lands, they have had to "do what had to be done," according to Antonio. "What had to be done" has included social activism, establishing local businesses and cooperatives, serving on county boards, and developing direct markets for their products.

Native New Mexicans, Antonio's grandfather raised sheep and Molly's father is a cattle rancher. After obtaining a degree in psychology from the University of New Mexico, Antonio returned home and married Molly. The couple attempted cattle ranching, but found it unrewarding; they then gravitated toward sheep. "Sheep are better producers," observes Antonio. They ranched at several places around the Chama Valley before inheriting the ranch where they live now. Antonio and Molly have worked side by side with their children, two of whom are now away at college "learning to think." Two employees and several sheep dogs round out the operation.

In talking about their life and accomplishments, Molly and Antonio are plainspoken and direct, refusing to glamorize the hard work of sustainable ranching and the obstacles they face. "We realize that this is part of a bigger struggle that is going on throughout the West. We know we are not isolated. But it is hard. It is tough to maintain the energy to keep it going. It is really, really hard."

Land ownership in this part of New Mexico is complicated, to say the least. Spanish land grants were established four centuries ago, and property rights and traditional uses became points of serious conflicts in the 1960s. Antonio was a teenager then, but remembers that his grandfather was passionate about the cause. This passion for remaining on ancestral lands and preserving the culture underlies Antonio's every action, although he modestly claims that he and Molly "are just trying to make a living." At times he has found it necessary to take a strong stand.

Several other factors currently threaten agriculture and ranching in the area. "Living in a beautiful place can be a curse," observes Antonio, pointing to second-home developments and gated communities that have sprouted up around nearby Lake Heron, a project of the San Juan Compact, which distributes water throughout the state. Other lands previously used for grazing were bought by The Nature Conservancy and sold to the state, and are now used primarily by hunters. And the Jicarilla Apache tribe has bought up several large ranches with revenues from gas and oil. "There is a certain justice to that," admits Antonio. Nevertheless, those pastures have been taken out of circulation for the locals.

Now the Manzanares graze their sheep near Tres Piedras in spring, then move them through the Carson National Forest to summer range near Canjilon, which can take almost a week. The sheep and camp have to be moved weekly to secure fresh pasture and conserve the land. Then the Manzanares bring the sheep to the ranch while the ewes are taken to the fall pasture and the lambs to Chama. Careful selection of grazing lands is necessary since Shepherd's Lamb is certified organic and grass-fed, and free of hormones and antibiotics. Their sustainable management system is based on traditional methods, and evolved using common sense, intuition, and some book learning. "We developed our own system of grazing, and of management, and of herding," says Antonio. "And it is a key thing that we still herd. So we still have

control. We have some control of where the flock goes, what it eats, how long it stays there, and how we use the country."

As the region's agricultural base has slowly slipped away, Antonio and Molly have looked for other ways to "keep the young people on the land." In 1989, along with Maria Varela and Rachel Brown, they founded Ganados del Valle, a nonprofit agricultural and economic development corporation whose mission was to ensure the continuation of sheep-herding as a way of life by making it economically viable.

The most prominent result of their efforts was the establishment of Tierra Wools, a thriving cooperative of weavers who craft beautiful handwoven tapestries, blankets, and apparel to be sold worldwide. Molly is a master weaver who has served on the board and as general manager. Much of the wool used comes from the Manzanares' own flock, providing an important market. Tierra Wools now supports seventeen worker-owners and employs twenty-five others. Walking through the shop, one is bombarded with earthy smells and rich textures and colors, many derived from natural dyes from local plants. And one is reminded of the enchanting beauty and rich tradition of northern New Mexico. The legacy of Ganados and Tierra Wools, notes Antonio, has been to teach skills to local people, allowing them to succeed in businesses and stay in the area.

With Lyle McNeal of Utah State University, Antonio and Ganados have also helped establish a breeding program to restore Navajo Churro sheep, a hardy breed suited to the Colorado Plateau. With nearly 40 percent of their flock now

Churros, the Manzanares have one of the largest Churro flocks in the country.

Developing and servicing markets for their sumptuous grass-fed lamb has become the most time-consuming aspect of the Manzanares' operation. In summer, Antonio and Molly's days are filled with a daunting number of activities. Yet servicing niche markets is the only way they believe they can survive amid the worldwide agribusiness market. They are members of the New Mexico Organic Livestock Co-op, a handful of producers who market to restaurants, specialty stores, and farmers markets throughout the state. They put close to 100,000 miles a year on their vehicles, transporting sheep to Zentiram Industries in El Rito, the only certified organic plant in New Mexico, as well as taking the meat to Los Alamos, Taos, and Santa Fe farmers markets weekly.

Antonio and Molly also maintain a direct-sales business with more than 1,300 customers. "People like to have a connection with the farmer and rancher," notes Antonio. "So much has come out about food safety issues. People like knowing where their food comes from and that it is raised organically." Of course, direct marketing means that record-keeping and book work consume many of his precious evening hours even as he remains actively engaged with the community.

Despite the backbreaking work and the toll on their health, changing land tenure, changing climate, government bureaucracy, uncertain markets, and struggles with powerful outside forces, Molly and Antonio think the life they've chosen is worth it. When asked what she considers their greatest accomplishment, Molly replies, "We have been able to raise four children on this land, and they've turned out OK." Antonio adds, "we are still here, and still together." Although they don't think of themselves as heroic, they are proud of their success in "making a living," and demonstrating a more sustainable way of life that benefits their community.

Right: The Manzanares' sheep furnish wool for the Tierra Wools weaving cooperative.

DOING WHAT HAS TO BE DONE

"JUST PARTICIPATE!"
The Babbitt Ranches

Coconino County, Arizona · By Rose Houk

Billy Cordasco

If you ask Billy Cordasco what sustainability is, he takes his time answering. "It's a hard concept," he says, "because the environment is always changing." The best explanation he's ever heard is "living off interest, not principle."

Billy is president of the board of Babbitt Ranches, a job he's held for nearly ten years. And though he may struggle with the word "sustainability," he has obviously spent a great deal of time thinking and experimenting with the idea. The Babbitt Ranch holdings and leases, close to 700,000 acres in northern Arizona, have been in the family for almost 120 years. They include the CO Bar and Cataract ranches, which stretch from south of the Grand Canyon all the way to the San Francisco Peaks and east to the Little Colorado River alongside Wupatki National Monument. They range from mid-elevation grassland, into pinyon-juniper woodland, to high-elevation ponderosa pine forests, among classic Southwest mesas and cinder cones.

In many ways Babbitt Ranches is a traditional outfit, running a cow-calf and yearling operation. "There's a routine built in," says Billy, "that's grown out of years of knowledge and learning." The Hereford cattle are driven to high country in summer and back to lower range in winter. Calves are born in the spring and branded. Fall is roundup time. Calves are taken to a ranch, raised another year, and then sold.

"We try to run cattle at 'drought levels,'" explains Billy, meaning that they run a minimal number of head based on the average carrying capacity of the range. In 2002, to weather the drought that has gripped the West for several years, Babbitt Ranches had to take the costly step of moving the cattle to several different pasture locations in the southern part of the state and even into Texas, holding them there until summer rains greened up the grama grass back in northern Arizona.

Like ranchers everywhere, Billy is concerned about the health of the grass, the water supply, and his cattle. But native wildlife, like pronghorn, also fascinates him. When he's out in his truck, he stops often, raises binoculars, and scans the endless platinum grasslands for the animals. He's concerned when he doesn't spot any, and rewarded when he does.

The larger view – evolutionary and ecological processes – also occupies Billy's thoughts and conversation. Among the most significant influences has been the "land ethic" expressed by the great conservationist Aldo Leopold. On the wall of his office is a poster that displays quotes from Leopold's classic book, *A Sand County Almanac* – albeit rearranged to reflect a landowner's priorities. Billy was elated when he got to go to Madison, Wisconsin, in 2003, near Leopold's Sand County farm. There, on behalf of the ranch owners, he accepted a wildlife stewardship award from the International Association of Fish and Wildlife Agencies for participation in grazing lease planning and implementation. The award cited Babbitt Ranches' efforts to provide water, restore habitat, and install wildlife-friendly fencing for native species.

Unquestionably, the most important person in Billy Cordasco's life was his grandfather, John Babbitt, who raised Billy after his parents died. The elder Babbitt, who ran the family's ranches for nearly fifty years, was a "man among men," says Billy. It was by his extraordinary example that Billy learned about ranching and other valuable lessons in life.

After graduating from Northern Arizona University with a business degree, Billy worked during the 1980s in several Babbitt enterprises, including a summer on the ranch surveying fence line. In 1992, at age thirty, he became president of the board. Billy Cordasco utters the word "blessed" many times as he talks about his life.

As a businessman, Billy is fully aware that he must answer to his board and shareholders. But whenever the board makes decisions about the ranch, he urges them to consider three things – economics, community, and environment. As he sums it up, "It's always about relationships," both natural and human.

During Billy's tenure as head of Babbitt Ranches, the company has pioneered a variety of projects that fall under the rubric of sustainability. A biological assessment of the company's land was completed and compiled into a publication. A new name was adopted – Coconino Plateau Natural Reserve Lands – which connotes the obligation and responsibility Babbitt Ranches holds. The company established the Environmental Monitoring and Assessment Foundation at Northern Arizona University as a conduit for landowners, agencies, and organizations to collect data and practice solid science on lands in the region. Babbitt Ranches also donated forty thousand acres of conservation easements to The Nature Conservancy and county, implemented holistic range management, and launched a watershed assessment project.

Billy is always looking for ways to raise and sell beef directly, rather than shipping off the cattle to distant feedlots to be fattened. Babbitt Ranches has produced beef jerky and hamburger patties, but to expand that endeavor, Billy notes, consumers must show their support by buying meat that has been raised organically and sustainably. A few years ago Babbitt Ranches also brought in bison, and "six girls – big ones" still roam the range, Billy notes.

"Ranching can be viable out here," Billy asserts, as he looks out on land that receives about nine inches of rain a year. If the environment were the only challenge he faced, Billy's job would be easier. But his task is far more complex – managing

Babbitt Ranches runs a traditional cattle operation – and strives to practice good land stewardship.

a huge landholding in a traditional family business, amid pressures of public issues and development that can quickly make cattle-raising less than economical.

So, rather than use the word "sustainable" to describe his land ethic, Billy Cordasco preaches the gospel of "just participate!" For him, the phrase embodies the belief that we can't manage or control land or the environment. To join, share, and be a part of ecological processes, he says, "we can only learn and understand."

Realizing Dreams under the Stars
Stargate Valley Farms

Holbrook, Arizona · By Roger Clark

To many people in their sixties, retirement means days filled with hitting golf balls, playing bridge with friends, or traveling the world. But Carol Poore and Dennis Swayda had another dream: to own a farm and raise goats.

The couple milk sixteen dairy goats by hand every twelve hours, seven days a week. "I spend four hours milking and six hours making cheese most every day," says Carol. "Stargate Valley Farms," Dennis explains, "is just getting started as our retirement business." That business currently includes raising purebred and American registered Toggenberg dairy goats and producing cheeses and organically grown vegetables that they sell at the farmers market in Holbrook, Arizona.

This energetic couple has worked for much of their lives in nutrition and organic farming. "We were lucky enough to be one of the first distributors for Nature's Sunshine encapsulated natural herb products," Dennis says. With business enormously successful, Carol and Dennis were able in 1992 to purchase their farm along the Little Colorado River, near Woodruff Butte seven miles southeast of Holbrook.

But the timing of their "retirement business" has had to be staged to meet competing commitments. They still speak at nutrition conferences, and have had to make the transition from their Phoenix-based home and businesses. "We finally

managed to move into a rental home in Holbrook," Dennis says, "and are planning on building a new home next to the farm where we keep our goats."

For the past several years, they have produced vegetables in the backyard of their home in town and in a neighbor's backyard. Located near the center of the small railroad and interstate town, their gardens appear as a verdant oasis amid paved streets and modest cinder-block homes. They produce a surprising amount and variety of vegetables, as well as peaches and apricots, from what amounts to less than a quarter of an acre of land behind the three homes. "I take advantage of the various growing niches created by sun and shade," Dennis explains. He points to peas, which do best in the partial shade of a fruit tree, and to peppers that thrive against a hot south-facing wall.

Their small greenhouse helps protect plants started from seed until after the last frost in mid-May. Dennis interplants a dense mixture of tomatoes, cilantro, salad greens, beans, carrots, and other vegetables that thrive on soil enriched by

Carol Poore

Dennis Swayda

goat and rabbit manure. "I'm a vegetarian, so we don't eat the rabbits," he adds. Into the soil go red worms and mulch from composted table scraps: "Nothing goes to waste around here." Drip lines and soaker hoses help to conserve water. Dennis actively rotates crops as they mature, so that a small plot might produce two or more harvests during the 120-day growing season. At peak periods of production, Dennis and Carol take 175 pounds of tomatoes to the weekly farmers market where they and a few other gardeners keep the townsfolk supplied with locally grown produce.

Beneath the shade of a large almond tree on the side of the house is a pen full of young goats, brown- and white-faced kids recently separated from their mothers. Carol explains that to protect the mothers' udders the kids are not allowed to nurse after birth, although she and Dennis continue to provide the mothers' milk to them. She releases two from the

pen. They immediately vie for her affection, until one finds the vegetable garden of greater interest. "They're a lot of fun – and smart, making them a handful to manage," Carol says. After coaxing the young doelings back to their pen, she notes that they only keep a couple of bucks for breeding. Out of earshot of Dennis, she quietly mentions that, in addition to supplying the Turquoise Room Restaurant in Winslow with goat cheese, she recently sold two young males to the chef for a special dinner event.

On the drive out to their farm, Dennis and Carol stop at a small canyon that cuts steeply through the red sandstone plateau on the south edge of their property. History comes alive here. Along the weather-darkened walls are petroglyphs pecked into the rock by people who farmed this valley some 900 years ago. Through binoculars, circular symbols, mountain sheep, and human-like figures are visible on the opposite wall. Sixty feet below is a small pool of water surrounded by the tracks of living animals. At the mouth of the canyon are blocks of sandstone rubble, the remnants of a dam abandoned by the Mormon farmers who settled the area more than a century ago. Carol and Dennis's property extends to near the base of Woodruff Butte a mile away. They regret that they could not prevent the adjacent landowner from mining gravel from the butte, much to the objection of the Hopi, who regard it as a sacred site.

Below the plateau, a complex of goat corrals, pens, feeding mangers, and milking barn stands next to an eighty-acre field where alfalfa was once grown. The property also came with two wells that fill a two-million-gallon earthen reservoir. Dennis, who is a professional dowser, will locate sites for other wells if necessary. They purchase animal feed made

Stargate Valley Farms' dairy goats graze on what the land provides but require milking twice daily, seven days a week.

from corn, oats, and barley produced by Mennonite farmers who guarantee that it is free of genetically modified plants. Carol recently returned from her former home in Iowa where she bought a used tractor and manure spreader. "It's hard to find used farm equipment in the Southwest because most of it winds up on farms south of the border," she says. Within the next year, they plan to begin growing vegetables and producing alfalfa for the goats here.

Dennis and Carol want to build their home up on the hill, where they will enjoy a view across the entire valley. When

the couple had to come up with a name under which to register their goats, Carol says the name "Stargate Valley" came to her when she looked up at the sky during a walk on a moonless night. "I'd never seen so many stars," she says.

Later she quips, "It took me twenty-six years and two husbands to realize my dream." Both Carol and Dennis feel they were destined to help restore farming to this ancient land. The only regret, Dennis adds, is "that I didn't discover this twenty years sooner."

IT TAKES A VILLAGE
The Diablo Trust

Coconino County, Arizona · By Sue and Tony Norris

Diablo Trust cowboys

The Prosser and the Metzger families have been neighbors for more than eighty years. Their combined ranches, the Bar T Bar and the Flying M, consist of 426,000 acres of federal, state, and private lands spread across a broad swath of northern Arizona southeast of Flagstaff, between Mormon Lake and the town of Winslow. This massive holding covers six different biological zones that range from ponderosa pine forest at 7,500 feet in elevation to Little Colorado River valley desert scrubland at 5,000 feet. It is home to a diverse wildlife population, including four threatened and endangered species. A defile, known as Diablo Canyon, separates the ranches from one another. It is from this limestone chasm that a unique collaboration took its name.

Formed in 1993, the Diablo Trust is now a flourishing collaborative land management team whose mission is to "maintain ranches as long-term, economically viable enterprises managed in harmony with the natural environment and the broader community." Besides the ranchers, participants include environmental activists, government range and wildlife officials, artists, academics and students, recreation enthusiasts and hunters, and any community member who wants to participate in the land stewardship process.

Seated on the weather-worn steps of an old cabin nestled high in the pines, Judy Prosser, who with her husband Bob and twin sons runs the Bar T Bar, recalls the group's beginnings. She explains that the climate in the early 1990s was more than a little hostile toward area ranchers. The Flagstaff newspaper was rife with "anti-grazing propaganda." Judy and her family wanted people to understand that not all ranchers do a bad job and so they joined with Jack Metzger's family to proactively tell their story. After all, both families had grown up, gone to school, and raised their families in the area. "We called a meeting of just about everyone we knew," says Judy, "invited a facilitator, and began the hard business of bridging the gap between ranchers and their critics."

Ten years and countless monthly meetings and team-building retreats later, the Diablo Trust has made considerable headway

Right: Pronghorn antelope are among the wildlife species of concern on Diablo Trust lands.

IT TAKES A VILLAGE

in bridging the gap between the various stakeholders' interests. "The collaborative allows us to work within a community, and the community to work with us," explains Jack.

As part of the collaboration, scientists from Northern Arizona University (NAU) and Prescott College have conducted research on comparative grazing practices, riparian and wetlands areas, piñon-juniper regeneration and distribution, and various wildlife studies. "The research helps us determine if we are getting the desired effects on the ground from our use, since there is a symbiotic relationship between grazing and the health of the land," observes Jack. Besides providing scientific feedback for future decision-making, these projects involve students and community members as hands-on participants. And the scientific community has come to appreciate the knowledge and know-how of those who live and work on the land.

During annual Diablo campouts held in conjunction with the Arizona Game and Fish Department, State Land Department, and Arizona Antelope Foundation, more than three miles of fence has been reconstructed to allow free movement of pronghorn antelope. Joint projects have been undertaken with a variety of organizations, including EcoResults, the Center for Sustainable Environments at NAU, and the Grand Canyon Trust. Presentations have been made to dozens of community groups ranging from the Kiwanis Club to the Ecological Society of America. A website, brochures, and a three-part educational video with curriculum tell the story of the Diablo Trust's goals and successes.

During their annual "Artists on the Land" days, visual and performing artists, writers, and musicians are invited to enjoy the beauty of the ranchland and to create songs, poems, and pictures in celebration. The results are then displayed and performed at the "Reflections of the Land" Art Show at NAU. Musician Trish Jahnke observes, "The best way to educate is through the heart. We know that only through heartfelt community can we bring about protection for our local family ranches." Jack fondly recalls the day he watched dancers with streamers fill a meadow with movement and color – somewhat to the consternation of the indigenous equine population.

In recognition of these unique collaborations, Diablo Trust has been designated a National Reinventing Government Laboratory, and has been honored by the federal Environmental Protection Agency. Still, such recognition does little to hasten the bureaucratic machinations that are holding up adoption of the comprehensive management plan developed in 1999. This plan includes all national forest, state, and private lands in the six vegetative management zones. Jack describes the environmental review of the Forest Service portion as "process paralysis," and counts it as a major frustration.

Other challenges that loom large include finding outside funds to finance improvements, as the ranches themselves have borne most of the cost so far; dealing with a burgeoning elk population that devastates cattle forage; and confronting continuing efforts to eliminate grazing on public lands.

When asked what progress has been made, the Metzgers and Prossers point to the restoration of several thousand acres of watershed through removal of invasive juniper trees. This has allowed native grasses to reestablish themselves, and has produced a richer and diverse plant community – thereby benefiting both cattle and wildlife. But all agree that the greatest accomplishment has been interpersonal – the building of trust between people of disparate interests and value systems in order to work toward the common goal of doing what's best for the land and the people who live and work on it.

The collaborative process has led Sierra Club activist Norm Wallen to this conclusion: "My experience with the Diablo Trust has convinced me that family ranching is the best hope for the rangeland of the Southwest. If they are driven out, I believe much of this land which I love will be either allowed to deteriorate through erosion and other symptoms of death or turned into senseless subdivisions."

Kit Metzger, Jack's sister, is in charge of the grazing plans for the ranches. When asked what she sees as the overriding purpose of their efforts, her answer concurs with Norm's: "to keep us on the land."

Staying on the land also means managing the economic risks associated with ranching in today's global economy. To this end, the families have worked to financially diversify their operations. Bob Prosser has had some success with a sod business. Jack Metzger is attempting to develop a market for small-diameter timber from the watershed restoration project, but explains that few businesses are willing to take a chance on a relatively small-scale supplier. Family ranch enterprises are between a rock and a hard place in other areas as well. They are too big to cater to local value-added meat markets, yet too small to process meat themselves. These two ranches have been successful in producing an above-average product on the open market, however, resulting in premium prices for their beef.

In the final analysis, it takes a little of everything to keep a family ranch viable: money, ingenuity, and hard work, as well as careful collaboration with government agencies, special interest groups, the scientific community, and the average citizen. The Metzgers, Prossers, and the Diablo Trust are showing that these days "it takes a village to raise a cow."

SECTION THREE
Wildcrafting

Wild plants and animals have sustained people in the Four Corners region for at least ten thousand years. And even in an era of supermarkets, harvesting products from the region's wild lands continues – carried on by practitioners who put a modern spin on this time-honored tradition. For these people, "wildcrafting" is a way of celebrating the bounty of local nature; of attuning one's senses to the rhythms of a particular, local place; and of placing a high value on the nourishing and healing properties of what nature provides.

One example is Phyllis Hogan, whose first contact with medicinal herbs came from an elder Spanish woman known as a *curandera*. Phyllis realized that much of this woman's valuable knowledge would be lost if it were not passed on – so she dedicated herself to documenting it. Now she runs a store, Winter Sun, where she shares this knowledge with others.

Even what most people consider common weeds can help sustain lives and livelihoods. Katrina Blair and her employees at Turtle Lake Refuge in Durango, Colorado, concoct gourmet meals from what grows around town, whether in wild places or in the untended spaces of urban lawns. Her "Dandelion Brigade" will come to your house, pluck out the unwanted plants from your grass, and use them as raw materials for nutritious foods.

Local knowledge is the common thread among wildcrafters. Dennis Arp manages 1,200 hives of bees from his home in Flagstaff. Each year he trucks his hives from low to high elevations, taking advantage of what's in bloom: desert mesquite trees in spring, mountain wildflowers in summer, then almond groves in California in winter. Each jar of his honey reveals in its color and flavor the distinctive place whence it came.

At Northern Arizona University foraging has become institutionalized. NAU's Center for Sustainable Environments may have the only university employees with job descriptions as "hunter-gatherers." Patty West and Teresa DeKoker of the Center's Community Wild Foraging Project gather purslane, horehound, prickly-pear fruit, and amaranth from farms, ranches, and public lands on the Colorado Plateau and sell it (recipe included) at the Flagstaff Community Market.

For all these foragers, a deepened connection to place, a reduced reliance on fossil fuels, and the health benefits that come of eating unprocessed, organic foods are among the many sound reasons to harvest natural "crops." If our society as a whole is ever going to truly embrace its immediate surroundings, then these pioneers are the vanguard.

REVERENCE AND RECIPROCITY
Phyllis Hogan's Winter Sun

Flagstaff, Arizona · By Ashley Rood and Rose Houk

Phyllis Hogan

Phyllis Hogan lives life according to her own script. Applied ethnobotanist-trader, mother-community leader, mentor-scholar, activist-musician – these are the many roles she assumes as part of her philosophy of reverence and reciprocity.

Phyllis's business, Winter Sun Trading Company, is in a historic building in downtown Flagstaff, Arizona. Winter Sun is a modern Southwest trading post, filled with treasures and tinctures collected from local artisans, wildcrafters, and healers. Phyllis is in constant motion, her long full braid and flowing skirt a blur as she simultaneously chats with customers and negotiates the small daily tribulations of a retail shop. In the front of Winter Sun, walls are hung with a beautiful collection of traditional Hopi *katsina* carvings, glass-topped cases are filled with silver jewelry, and yucca baskets hold bundles of sagebrush. In the back, a smaller, low-ceilinged room – filled with herbs, spices, teas, soaps, and salves – has the feel of a century-old apothecary shop.

The impetus for Phyllis's life's work came more than thirty years ago, when she made her first trip to the Hopi Reservation in 1971. She became close friends with Herbert, a Hopi medicine man, and read Alfred Whiting's *Ethnobotany of the Hopi*. In correspondence, Whiting suggested she undertake a study comparing Navajo and Hopi plant-gathering strategies. She began a long process of teaching herself about ethnobotany.

Then, for the next four years, she spent time with her two young daughters in tow "just looking around" the Sonoran Desert, meeting like-minded people, and at times delivering herbs to the Hopi. Disillusioned with conventional medicine, she was searching for natural alternatives for her daughters' usual childhood ailments. That search led her to a small Mexican curio and traditional herb store in Coolidge, Arizona, owned by Señora Marion Valencia. One day Phyllis waited patiently at the screen door with her two daughters in hand. Señora Valencia was an herbalist and *curandera*, or traditional healer, and she was wary of strangers. But she opened the door out of curiosity. At that moment a long friendship began.

As Phyllis watched and learned about medicinal plants from Señora Valencia, she wondered who would document this traditional knowledge. Her conviction that she could do so led her to open her own store first in Coolidge in 1976, then in Flagstaff in 1978, and a year later on Route 66 in downtown Flagstaff. Her contact with the Navajo began here, especially with medicine man Sam Boone, Sr. "It all spiraled from there," she recalls.

Phyllis approaches ethnobotany from two angles: practice and reciprocity. Applied ethnobotany is the practice end of things. "I wanted to document the plants and their uses," she says, "and the only way to do that is to use plants and get involved in them." Reciprocity takes place in her Flagstaff store, where the trust between her and Native American and Hispanic people has developed, and where she offers to the community an exceptional selection of medicinal herbs.

This combination of practice and reciprocity has taken Phyllis down a unique path. She has been granted the rare honor of being invited into indigenous communities, and of having people from those communities come to her as well. Collecting herbs with native people, she has seen how they respect the land, and in turn how respect must be accorded the plants themselves. "I learned never to take anything without respect and asking," she notes. When she gathers, she observes an age-old tradition of honoring the plant by making the proper

A sample of Winter Sun's line of herbal products.

prayers and offering pollen, cornmeal, or tobacco. She notes that the practice also requires proper preparation, such as dressing appropriately and disregarding worldly concerns.

The places where Phyllis gathers growing things are to her "wild gardens," some of which have been tended for hundreds of years by particular clans. Many of them are in or near towns or cities. The places where the plants grow are as important as the plants themselves, and the sites are carefully guarded because they are sacred. "That's what is being lost," Phyllis asserts, "because the information about the significance of the sites is not being passed on to new generations." She expresses a profound sense of responsibility toward the knowledge she's been granted. And even though she is non-Indian, she's trusted "because I've been in place so long. That's what reciprocity is."

Through the years, Phyllis has seen her profession gain respect in wider circles; it's now nearly mainstream, not a cause out on the fringe. She expresses pride that she has remained independent in her research, and has had the persistence to prove that there is stability and honor in the work. Her measure of success after so many years is straightforward. She explains, "The people I started working with thirty years ago are still my friends, and their grandchildren now are listening to my stories."

In addition to her full-time responsibilities as proprietor of Winter Sun, Phyllis, with Michael Moore and the Sam Boone, Sr. family, founded the Arizona Ethnobotanical Research Association. The nonprofit association was established in 1983 to document and preserve traditional plant knowledge. It has expanded over the past two decades to include multicultural and bilingual regional education programs, an heirloom seed bank, and an annual conference.

Phyllis is always on the go. She is advising the new Black Mesa Water Coalition, building her own straw-bale home, and teaching at the Southwest School of Botanical Medicine. Perhaps most important, she has passed down her passion for traditional medicine and community to her two daughters, DeeAnn and Denise Tracy. The sisters are both graduates of the Southwest School of Botanical Medicine and each has created her own original, all-natural skin care line, Peak Scents and Super Salve.

Amid the chaos of Phyllis's many ventures is this constant: she thrives on community. Her philosophy can be wrapped up in the term "bioregionalism," which connects a range of community conceptions from the complexity of its plants to scientific appreciation, experiential understanding, reverence, and most important, the act of reciprocity. She extends this philosophy into her work and her life, rewriting the script joyfully and respectfully every step of the way.

Phyllis's Winter Sun is a modern-day trading post with traditional products.

Alchemy in the Kitchen, Fire on the Land
Turtle Lake Refuge

Durango, Colorado · By Rachel Turiel Hinds

In Durango, Colorado, a small revolution is taking place. It simmers quietly in the backyards of people who harvest dandelions, amaranth, and other wild weeds for dinner salads. It gathers steam in a historic district alley where up to one hundred people each month eat wild, locally harvested, raw-food lunches. And it extends to the nearby woods, where people pick and use the wild berries of summer – choke-cherries, serviceberries, hawthorn. This revolution is on fire.

Katrina Blair, Durango native and the founder and visionary of Turtle Lake Refuge, is the living root of this revolution. Since 1998, Turtle Lake Refuge has been teaching people the benefits of eating wild-harvested, locally grown, and "living" (uncooked) foods, all of which, Refuge workers believe, decrease stress on our bodies and the Earth.

Katrina's work, life, and passions are so intertwined that it is impossible to tell where work ends and the rest of life starts. A fall workday includes climbing an apple tree and picking bushels of apples for Turtle Lake benefit lunches, served twice a week. In spring, tender dandelion leaves are plucked from mountain meadows and backyard weed patches for pesto, salad, or "Wild Mint Magic" bars. Katrina loves to harvest

Katrina Blair

food. Just the sight of her freezer's contents – gallon jars full of chokecherries, fresh-pressed apple juice, serviceberries, and a small, coveted bag of buffaloberries, all harvested locally – make her swoon.

The mission of Turtle Lake Refuge is to celebrate the connection between personal health and wild lands. The nonprofit aims to create a more sustainable community by linking the value of a healthy internal environment – our bodies, with a healthy external environment – the Earth. All the profits from its varied fundraising activities go toward promoting sustainable living practices and preserving open space.

Why local, wild, and raw? "Locally grown foods minimize the resources required for transporting our food from faraway," Katrina explains. "This can mean buying from local farmers, picking a neighbor's unwanted apples, or growing food in our

A wild-foraged lunch at Turtle Lake Refuge.

eat entirely raw one day each week to take the strain off their digestive systems and promote general health. Doctors said Pat would be in a wheelchair by forty. Instead she is free of arthritis. Today many people come to the Refuge because of chronic health issues, including poor digestion and fatigue.

Like the swankest big-city restaurants, Turtle Lake Refuge has no menu, simply one special of the day. The lunch clientele is as eclectic as the food. Young, pierced, patched, and dreadlocked twenty-somethings break flax-seed crackers with businesswomen and grandfathers. Topics among the patrons range from college classes to menopause.

This Friday, Sissy Mueller serves up veggie burgers, sweet potato fries, onion rings, green salad, and a slice of apple pie for dessert. It's all raw, and thoroughly delicious. "It's the All-American meal 'cept good for you," Sissy laughs as she sets meals on a table. She and the other chefs are generous with their recipes. There is no secret sauce, though exact proportions are vague and directions often dictate "a pinch of this" or "a bunch of that."

Lunch is determined by what's in season. When wild watercress shoots up in spring, it becomes the cornerstone of the daily green salad. In fall, expect local apples, plum, grapes, and pears. At the peak of summer's richness, anything is possible. Turtle Lake Refuge draws from many local gardens in the summer, and from wild land as well. In the colder seasons salads come almost exclusively from local greenhouses. The five employees and three interns (who receive college credit) spend much of the summer harvesting wild foods and preserving them for the winter. Berries are frozen, juiced, or

own backyards. Wild foods connect us to the land we inhabit and provide one of the highest mineral sources available. At our lunches we include something wild every day: a wild weed salad, hawthorn berry pie, or yucca fruit salsa, for example."

Her mother, Pat Blair, who was diagnosed with arthritis at age seventeen, introduced Katrina to raw food. The family would dried into fruit leather – foods frozen or dehydrated at less than 120 degrees retain their living enzymes. Large quantities of dandelion, amaranth, and mallow – a bane of gardeners, but highly nutritious – are dried, powdered, and added to the popular "Wild Mint Magic" bars.

The latest in this evolving vision of many creative minds is the Dandelion Brigade, a pool of laborers who visit homes and dig up dandelions, roots and all, for less than it costs to spray a lawn with pesticides. Not only will the lawn be free of chemicals and immediately safe for children, dogs, and bare feet, but the Dandelion Brigade will teach you how to use dandelions for food and medicine. They will bring a bicycle-powered juicer to make fresh dandelion juice as they pick; for an extra charge they will make dandelion wine or beer from the flower heads and bring the homeowner the finished product.

Katrina's skin is sun-kissed and her body shaped by the work that is simply her life. Strong, ropy hands chop apples. Sinewy legs that have climbed many mountains dart from kitchen to pantry gathering up supplies for chokecherry macaroons. For the past four years Katrina has been teaching a four-week class called "Chi Foods." Students learn to sprout seeds at home and create delicious, gourmet meals out of living foods. In summer, Katrina puts a twist on it and offers "Chi Foods – on the Wild Side," in which people learn to identify and prepare edible wild plants in the field. Does she fear that exposing so many people to wild foods will result in their depletion? "No," she says, explaining that 90 percent of the plants she uses are wild weeds that are in no danger of being over-harvested.

"What happens in agriculture today is a farmer plants one crop after spraying the land with herbicide to remove all competing plants," Katrina says. "Not only does this deplete nutrients and life from the soil, but it decreases food for wild animals and pollinators. When we plant our backyard gardens and farms organically, we can harvest the weeds that grow without additional water, fertilizer, or care, as well as our cultivated crops. Plus, wild foods have the highest nutrients so you don't need to eat as much. I think the more we teach people about wild foods the more we can sustain them and achieve a higher quality for life for everyone."

After getting a batch of chokecherry macaroons in the dehydrator, Katrina discovers that someone soaked a gallon of oat groats thinking they were wheat berries, which need to be sprouted for wheatgrass. "Oat groats don't sprout, so I guess we won't have wheatgrass for tomorrow – but we could make cookies out of these," Katrina suggests. Another employee offers, "How about carob hawthorn berry cookies?" "Yeah, with fresh mint," someone else suggests. And a new recipe is born.

This is how life works at Turtle Lake Refuge: when the floor drain overflows with water, it's a reminder to mop the floor; when a basket of cookies is jostled at the farmers market and each one breaks, you have samples to share. When your life, work, and passions are all the same, and you have a burning desire to share them all, people pay attention. Word spreads, enthusiasm swells, and a small revolution grows.

Weaving Tradition and Change
The Basket Makers of the San Juan Paiute

Willow Springs and Piute Canyon, Arizona · By Roger Clark

In the 1950s, when Bill Beaver began his career as a trader to the Navajo in Shonto, Arizona, the San Juan Paiute people were also living in that part of northeast Arizona, farming and raising livestock. Once nomadic traders and hunters, they traditionally used as many as 175 different plants for a wide variety of functions. One of the primary uses was basketry. With one type of basket they collected and winnowed seeds. They crafted larger baskets of yucca, willow, sumac, and rabbitbrush, and used them for storing or carrying food. They sealed bottle-shaped baskets with pitch for carrying water, and placed heated rocks in tightly woven "boiling baskets" to cook stew.

As they settled down to farm, the San Juan Paiute no longer needed utilitarian baskets; instead, they turned to weaving baskets that the Navajo valued for wedding ceremonies and to making decorative baskets collected by tourists and traders. Their skill in adapting to changing sources of sustenance continues today as a tradition reflected in their changing styles of basketry.

Bill bought and sold these baskets, and added some of the more unusual and "old-timer" baskets to his personal collection. He collected newer, "plaque style" baskets with multicolored animal figurines, as well as the loosely woven "seed beaters" made for collecting seeds. Bill's growing interest and trade in Paiute basketry continued when he

moved a hundred miles away to become the proprietor of Sacred Mountain Trading Post, north of Flagstaff. His relationship with the Paiute women who made the baskets continued to prosper. But cash flow problems led him to sell his collection of nearly a hundred baskets to a collector who subsequently sold it to the National Museum of Ethnology in Osaka, Japan.

Bill reflects on what happened next: "I talked to the women about putting a second collection together. They showed great interest. Then I loaned them a copy of publications about Southern Paiute ethnohistory that included photographs of utilitarian baskets taken during John Wesley Powell's 1872 expedition. And the response was great." Pretty soon beautiful baskets started showing up, and Bill noted who made them and what they had to say about the basket. He explains, "Many times they'd say, 'I got it out of those books you gave us.'" The women began bringing in baskets with their own designs. They were reconstituting the old at the same time they were stimulating new ideas.

In 1983, museum curator Bud Whiteford visited Bill at the Sacred Mountain Trading Post. As noted in a catalogue from

Right: Grace Lehi, Mabel Lehi, and Rose Ann Whiskers have passed the art of Paiute basket making from mother to daughter.

WEAVING TRADITION AND CHANGE

an exhibit at Santa Fe's Wheelwright Museum, Whiteford "found an unexpected treasure trove of contemporary baskets produced by a relatively unknown group of Native Americans, the San Juan Paiutes…. These initial encounters with the San Juan Paiute basketry collection at Sacred Mountain, and the associated information provided by trading post owner, William Beaver, have culminated in the exhibition at the Wheelwright and in the publication of this catalogue." The resulting exhibition, *Translating Tradition: Basketry Arts of the San Juan Paiutes*, was on display for nearly five months during the winter of 1986.

Mabel Lehi, whose work was featured prominently in the exhibit, still gathers native plants to make baskets and is continuing to teach the next generation of Paiute basket makers. Marie Lehi, her mother, was eighty-four at the time of the exhibit, and was quoted as saying, "The first person I learned [basketry] from was my mother. I spent a lot of time splitting sumacs before I ever made a basket. My first basket was sure ugly. But I got more skillful. I can remember when my people used 'carry baskets' for getting foods and I can remember how to make all these kinds of food baskets. I have taught all my daughters and some grandchildren." Marie died the evening after she was honored in 1992 at the Sedona Arts Center, where she received the "Arizona Living Treasures Award."

In a visit with Marie's three living daughters in Tuba City, Mabel jokes about their time spent gambling in the casino after a healthcare conference they attended. Bill asks, "So how'd you do?" "Phutt," replies Grace Lehi, as she slaps her hands together and then opens them like an escaping butterfly. Bill notes that Grace understands English but mostly speaks Paiute. Behind her on the wall is a photograph of an award-winning coiled tray with butterfly designs that Rose Ann submitted for judging at the Museum of Northern Arizona's Festival of Pai Arts. "She sold the tray for a good price," Mabel recalls.

Today, about eighty-five people are registered members of the San Juan Paiute tribe. Official recognition of the tribe and the creation of two small reservation islands surrounded by the Navajo Nation came belatedly, in 1998. During negotiations to resolve the long-standing land dispute between Hopi and Navajo, the San Juan Paiute managed to assert their legitimate claims to land at Willow Springs west of Tuba City and at Piute Canyon near Navajo Mountain.

Despite change, the tradition of basket making is being passed from one generation of San Juan Paiute to the next. Bill cites Rose Ann Whiskers as a prominent example of the next generation, as well as the King and Owl families from the Navajo Mountain area. "I'm on my fourth collection of Paiute basketry," he says. "In each one, there are examples of new styles and emerging artists."

Bill also notes the change that is occurring in who is making baskets. "Through the years I've seen men producing more and more work that was traditionally done by women," he says. Richard Graymountain, for example, has been selling baskets to Bill since he was a teenager. They're no longer essential to life in quite the same way, but baskets continue to reflect the changing lives, and the creative impulses, of the San Juan Paiute people.

WHERE TRADITION MEETS QUALITY
Zuni Furniture Enterprises

Zuni, New Mexico · By David Seibert

Zuni Furniture Enterprises has a slightly different take on the bottom line than a typical business. For this company, while economic concerns are important, cultural principles also guide the work. Zuni Furniture Enterprises, located at Zuni Pueblo in northwest New Mexico, has been a contributing part of the local economy since the early 1990s. Employees at the company craft locally harvested wood into fine, hand-carved furniture. The wood comes from tree thinning projects in the area. This tree thinning reduces fuel buildup in the forests that can otherwise contribute to catastrophic forest fires.

Production and output aren't the only priority at Zuni Furniture. Instead, says manager Sterling Tipton, local artists hand-paint culturally significant designs on the custom furniture in traditional colors schemes approved by the tribal council. Pale pine chairs, chests, and cabinets glow with brightly painted, eye-catching designs that resonate in Zuni culture. Each piece is signed and numbered by the artists. The personal investment and detail in every carefully crafted piece are reflected in the organization's motto: "Where Tradition Meets Quality."

The company strives to be smart with its resources. Wood scraps end up in the new BioMax 15 heating unit. The BioMax is an experimental "bio-powered" machine that transforms dry wood chips and other organic waste material into useful heat and electricity. This not only reduces overall waste for Zuni Furniture Enterprises, but provides the company with low-cost energy. The BioMax, on loan for testing from Colorado-based Community Power Corporation, can raise temperatures in the production building twenty degrees in about two hours. This considerably shortens drying times, especially in the winter when longer drying times can slow production to 40 percent capacity.

Zuni artist paints traditional motifs on a bench.

Perhaps the most important part of the Zuni Furniture Enterprises success story is the merging of cultural and economic concerns into a viable business. When Sterling Tipton, a native of Zuni, came to the job a few years ago, he brought with him ideas gleaned from years of international consulting work. Past approaches, grants, and local interest had been crucial to the enterprise in the past, but, as Sterling puts it, "one thing they couldn't buy was business experience."

Sterling is quick to note that Zuni Furniture Enterprises had been a locally run, functioning furniture producer before he arrived. But a pattern of government grants, too much emphasis on niche markets and tourist items, and a lack of personal investment hampered efforts at development and independence. "This is not a program," Sterling notes. "It's a business." Typical grant-funded programs can be useful, but the performance of the business and the individual employees sometimes suffer, as do quality and sustainability. As many observers have noted, programs and grants end eventually. The cycle of applying, receiving grants, and always looking to the next funding source creates a climate of instability and continuous dependence. Sterling would have none of it. He made immediate changes – and saw immediate results.

Some of those changes were details – albeit important details. "You got to protect your employees," Sterling notes, as he points to the fans that pull paint fumes out of the work area. A new paint room and extensive ventilation systems are only two of the simple but vital measures that at once improve both workplace quality and productivity.

Left and above: Furniture produced at Zuni Enterprises uses local woods and culturally appropriate designs and colors.

"Much of the problem in linking the business and cultural worlds arises from misunderstandings of local cultural ways," Sterling explains, "but this does not need to be the case." He plans to continue this work for the long term and has inspired others to believe in his philosophy. The company already has outlets in Tennessee, Seattle, and Albuquerque, and it plans to enter the market at Santa Fe – but not with tourist items. Products for tourists are typically seasonal, small-scale, and appeal to a market that is too often "here and gone." Like grants that can build dependency rather than self-sufficiency, the tourist market can simply "teach people to jump through hoops," says Sterling.

Sterling strives for a process and products that combine economic success and cultural traditions and values. While the cultural dimensions of this business remain foremost in his mind at all times, his approach is to design and operate "the best furniture company – period." At the same time, he wants people to acknowledge that the operation is "native owned," and that these domains need not remain separated, either in the minds of Zunis or outsiders. As Sterling says with a smile, "It's about economic development right here – and I'm gonna be here for a while."

IMPARTING HOPE FOR THE FUTURE
Melissa Porter's Wildcrafted Herbs

Chama, New Mexico · By Rosemary Logan and Peter Friederici

Bob and Melissa Porter

As you come over the hills into Chama, the land takes on a fresh, spring green color. It's not just the irrigated pastures of alfalfa along the river, but also the oaks and aspens and grasses that cover the steep slopes around the village. At nearly 8,000 feet in elevation, Chama is a well-watered highland oasis and a splendid place to grow a wide variety of herbs and other medicinal plants.

It's a long way from Santa Barbara, but still this small northern New Mexico town is an unsurprising destination for Melissa and Bob Porter, former Californians who a few decades ago were looking for a place to raise their children close to the land. They had enjoyed living near the beach on the California coast – Bob was an avid surfer – but gradually found the place too expensive, too crowded, too confining. They disembarked in Chama, a long way from the ocean, and haven't had any regrets.

Chama has only about 2,000 residents. It takes creativity and ingenuity to make a living here. When the Porters first arrived Bob got a job thinning trees for the Forest Service, and the couple used some of those logs to build the first part of their house – a log cabin – on six acres a couple miles outside town. They had two children, and worked odd jobs for years to pay for the house and the additions they later put on. Melissa was a librarian in town, and Bob ran a trash collection service. They raised goats and made cheese and milk for a while. Bob helped other people build houses and learned how to make furniture for sale.

Bob and Melissa learned how to keep bees. They began selling fresh produce and canned jams and jellies at local markets. They discovered many of the tricks required to garden successfully at high elevation: using row covers to protect tender plants from frost, placing shade covers over leafy greens to shield them from strong sunlight, applying chopped dead flowers and grasses as mulch, even keeping trays full of new seedlings warm in the winter by placing them atop their water bed (during those weeks the couple sleeps in another bed). They added on to the house a few times; it now includes a sun room/greenhouse, a second story,

Right: Melissa both collects wild herbs and cultivates many favorites on the farm.

IMPARTING HOPE FOR THE FUTURE

and a stone addition. In winter they rent cross-country skis and boots to area visitors.

A significant part of their livelihood, though, is Melissa's wildcrafting business. She collects and dries about fifty types of plants and sells these to a couple of stores and at a farmers market, as well as at the ski shop. It's a passion that began back in California. "It all started in 1970," Melissa explains. "A friend pointed out some lambsquarters. I was in awe. Here was this weed that was so common, and so delicious."

Once in New Mexico, Melissa began recognizing the local plants one at a time, without formal instruction. She learned to identify and collect everything from mint to mullein to yarrow, from snake broom to rose hips to echinacea. She and Bob developed a yearly collection cycle that begins in May with wild parsley, peaks during the summer rainy season in July and August, and tapers off in October when they dig for osha roots up in the mountains. She also began to plant her favorites close by.

Today the Porters' house is surrounded by thriving gardens full of herbs and flowers where Melissa does much of her collecting. "I gather the plants by hand," she says, "then tie them in bundles and let them dry. Things dry fast around here. Then I strip the leaves off the stalk and store them in glass jars in a cool, dark, dry area. A lot of the herbs you buy in the store have been cut and sifted, but I try to keep them intact. The more you handle them, the more you lose." Because glass jars are expensive she sells her wares in plastic bags, but encourages customers to transfer the herbs back to jars whenever possible for better preservation. Many herbs

have strong medicinal properties and may interact with prescription drugs. For that reason, Melissa doesn't prescribe uses for the herbs she gathers, but rather encourages people to do their own research about their workings.

Melissa shares her knowledge with others by conducting identification walks for a variety of audiences, from girl scouts to tourists to local folks. "Some are people from the community who never paid attention until their grandma passed away, and then they noticed there was a lot they never learned," she says. "Some are health professionals. Some are kids. One woman just wanted to know what was growing in her yard, and she was amazed. It was just a little yard, but there were so many plants there that she could eat or make tea from. People get excited about this."

For Melissa, wildcrafting carries a spiritual aspect that is very much in keeping with old traditions of stewardship. She likes the work of gathering and drying herbs, but she also likes knowing that she is contributing both to her community and to her place. "It's part of sharing," she says. "The human family is here to take care of the earth. It's not just gathering and selling, it's a whole educational process, teaching the responsibility to take care of the planet, teaching respect for the plants and for each other, and imparting the hope that the Bible imparts for the future." The quiet town of Chama, and the practice of wildcrafting, are for her the living embodiment of one of her favorite verses from Psalms 37:11: "But the meek ones themselves will possess the earth, and they will indeed find their exquisite delight in the abundance of peace."

THE SWEET LIFE
Dennis Arp's Mountain Top Honey

Mountainaire, Arizona · By Peter Friederici

When he was a teenager growing up on an Iowa farm, Dennis Arp once went squirrel hunting with friends and came across some beehives. The boys decided they'd try to extract the honey. Dennis borrowed a veil from his dad, who'd kept bees in the past. Lacking experience, the youths didn't quite master the technique of taking honey from the hives of bees that didn't want to give it up. "Pretty soon there were bees inside the veil," Dennis says, "and we decided it wasn't worth it."

Dennis has had a bit more practice by now. He started keeping bees as a hobby at the suggestion of a couple of coworkers about twenty-five years ago, and now, from his home in Mountainaire, just south of Flagstaff, Dennis manages an enterprise of 1,200 beehives that on average produce 100,000 pounds of honey a year. As proprietor of Mountain Top Honey, he's a successful entrepreneur whose work is firmly centered on the Colorado Plateau, even as he sometimes sells his products in places as far-flung as Taiwan.

Managing bees is an exercise in timing and motion, especially for someone who lives at 7,000 feet where nothing is in bloom for at least six months of the year. As a result, Dennis and his part-time assistant, Tom Hedwall, are on the road a great deal. At the beginning of February they travel to California to place beehives in the almond groves. Almond nectar doesn't make good honey, but almond trees must be pollinated by bees, and the growers pay well for that service. By mid-March it's time to move the bees back to Arizona and the dwindling orange groves of the greater Phoenix area. The orange blossom honey the bees produce there is light and tangy and popular with visitors and residents alike.

Tom Hedwall

In May and early June, Dennis and Tom move the hives to the desert outskirts surrounding Phoenix, where the bees forage on catclaw, mesquite, and paloverde flowers. Sometimes the two men have help from Dennis's son, who is in college. Always they work long hours. The beehives are moved only at night, when the bees are resting. In the desert it's hot even at midnight. "We spend a whole lot of time in the middle of the night moving bees, then working all the rest of the day too," Dennis says. They drive a flatbed truck. The wooden hives are stacked on pallets and moved with a forklift. In one night they can move 120 hives – a tenth of his total.

It takes long hours and many miles to maintain more than a thousand bee hives.

This summer the bees out by A-1 Mountain produced little honey, while those thirty miles to the southeast on Anderson Mesa made a bumper crop – a hundred pounds per hive. In November, after the bloom, Dennis and Tom extract the honey from the hives. They work in the "honey house" – a building the size of a two-car garage next to Dennis's house. The yard is full of fifty-five-gallon drums and five-gallon buckets, mounds of yellow wax, and machinery of indeterminate purpose.

The honey house is crammed with hives and with the machinery needed to extract and process the honey. "All this stuff makes my electric meter spin pretty good," acknowledges Dennis. Each hive consists of stacked white wooden boxes, each of which is lined with an array of rectangular wooden frames in which the bees build their combs and make their honey. To extract the honey, Tom places the frames on a pair of moving chains that carries them past a vibrating knife, which cuts the wax caps off the combs. Wax and honey slide gloppily down into a vat. A sweet, earthy smell fills the room. The frames continue moving on the chains into the "extractor," a large stainless steel vat in which they're spun at high speed. The centrifugal force presses out the honey. It and the leavings from the uncapper end up in a sump tank built into the floor, from which the honey is pumped into two holding tanks that each can accommodate almost two tons of honey. The leavings go into a small, square machine in which a heater separates the remnant honey from the wax.

By early summer the bees are moved to the Little Colorado River valley near Winslow and Holbrook, where the honey comes from camelthorn. This is a thorny and noxious weed, but it produces fine honey. When the summer rains start in July, it's time to move the hives close to home, to the grasslands and ponderosa forests around Flagstaff. Where specifically to move them is a roll of the dice. The monsoon rains are irregular. "Sometimes," Dennis says, "there might be flowers everywhere in one place, and you could go a mile one way or the other and not see any flowers. It can be kind of hit or miss." By the time a good bloom happens, it can be too late to move many hives there. Like a wise investor, he finds it prudent to split up his hives in various locations.

It's a process that looks messy, but in fact little is wasted. Dennis sells the beeswax to candle and salve makers, the pollen at health food stores, and the resin-like propolis, which the bees use to make repairs and which has antibacterial qualities, to makers of herbal medical products. He sells much of the honey directly to customers at a variety of stores and farm stands in northern and central Arizona; some goes to local restaurants; some is sold to a bakery in the Phoenix area; and then there is the occasional inquiry from Taiwan or elsewhere.

After helping Tom with the extraction process, Dennis sits by a row of five heated tanks at one end of the room. It's here that the honey ends up, in its diverse flavors and colors – orange blossom, desert wildflower, mesquite, camelthorn, Flagstaff wildflower – each heated to 130 degrees so that it can flow smoothly into glass jars. At one point he samples a bit of the honey that came from the ponderosa forest near Flagstaff this summer. It's tangy, almost a bit acidic, and very different from mesquite honey (dark, almost like molasses) and camelthorn (very light).

"It has a really nice, fresh flavor, doesn't it?" he says. "At an industrial scale you really lose that individual flavor and character. Most supermarket honey is mixed from honey from all over – Argentina, China, Mexico. You don't know where it's from."

That's far from the case with Mountain Top Honey. Dennis fills jar after jar from the barrels, labels them to indicate precisely what's inside, and places them in boxes that he'll drive himself to a local store. It is hard, grinding work to get the honey to this point, but as the accumulated sunshine of spring and summer, of goldeneye and rabbitbrush and buckwheat and fleabane, flows into the clear jars and glows golden under the fluorescent lights, it is also easy to be sure that it is sweetly, deliciously worthwhile.

Each jar of honey captures summers sunshine.

MODERN-DAY HUNTER-GATHERERS

The Community Wild Foraging Project

Flagstaff, Arizona · By Gary Paul Nabhan and Rose Houk

Teresa DeKoker

Patty West and Teresa DeKoker may be the only university employees in the United States to boast job descriptions as "hunter-gatherers." As coordinators of the country's first community wild foraging project, they gather wild foods from farms, ranches, and public lands on the Colorado Plateau, and assess the productivity and sustainability of harvesting such foods from different patches.

Their work does not stop there, however. The two are also documenting the best ways to clean, preserve, and prepare these foods, along with finding the most effective means of marketing their harvest.

Based at Northern Arizona University (NAU) in Flagstaff, the project has a stated mission: "to support regional collectors, farmers, and foragers through a collector-consumer organization... to build enduring relationships that will nourish us with great-tasting, regionally collected, and sustainably harvested food."

Patty West puts it in a more personal way: "Our project began almost as a dream in the winter of 2002. I realized I wanted to use my botany skills for more than esoteric research. I wanted to be involved in something that would benefit our community."

With staff from NAU's Center for Sustainable Environments, Patty came up with a plan to enroll Flagstaff residents as "member-shareholders" in the project. With the help of volunteers, she and Teresa roam the riparian corridors of the Verde Valley, the piñon-juniper woodlands of the mesas, and the ponderosa pine forests of the San Francisco Peaks region to locate harvestable stands of delicious, nutritious wild foods. They are always discovering innovative ways to gather and process them, and also put out a regular newsletter documenting their trials as well as their successes in finding and fixing wild foods.

Nature's greens and fruits provide a cornucopia of food and drink. Depending on the season, something is almost always ready for picking in the region's deserts and mountains. Foraged products include blackberry and grape leaves, plantain, yucca blossoms, pine needles, and horehound; cholla

Right: Harvesting prickly-pear fruits requires a careful touch.

MODERN-DAY HUNTER-GATHERERS

cactus buds, prickly-pear fruit, purslane, lambsquarters, mesquite beans, and amaranth; and acorns, walnuts, crabapples, and rosehips.

Teaming up with master chef Francisco Perez, Patty and Teresa held a "coming out" event in the summer of 2003 to introduce the Flagstaff community to some sophisticated dishes using locally gathered foods. The menu featured stuffed grape leaves, empañadas, pestos, and wild spinach pies. More than 200 curious folks sampled the feast and were won over by the extraordinary flavors and fragrances of the region's native edible plants. In that summer the wild foraging project was able to tally twenty-one member households who were delighted to

Jojoba nuts are among the specialty foods collected by the Community Wild Foraging Project.

receive six to eight wild foodstuff deliveries each month as the bounty of the season unfolded.

Although the foods come from wild (undomesticated) native and naturalized plants, they are collected most often from "working" landscapes – the agricultural fields, hedgerows, pasturelands, orchards, and woodlots that form a large percentage of the plateau's rural landscapes. Patty and Teresa are excited to be introduced to an organic orchardist in the Verde Valley who is thrilled to have them take all the purslane they want from between the rows. Otherwise, he'd be hoeing it out himself on a hot summer day. Private property furnishes a good harvest of prickly-pear cactus pads, and a few months later cactus fruits are collected on state land. In the fall, foragers gain permission from a private landowner to gather mesquite pods. A few trees provide pounds of beans that can be dried and milled into a protein-rich flour, which has been a staple of the diets of southwestern peoples for thousands of years.

Foraging undeniably requires a commitment to lots of manual labor. It often means down-on-hands-and-knees positions, pulling, clipping, and carefully avoiding thorns and spines. Because some of the harvested parts are perishable, a method of cooling is required. And with some financial risk involved, having subscribers pay ahead of time means the economic risk of the foraging enterprise is spread out, Patty says. The pilot project got off the ground with outside funding, but the ultimate hope, she adds, is to create an ongoing business.

Foraging wild foods carries a broader responsibility as well. Certain rules are observed, Patty explains, to assure sustainability of natural "crops." Foragers gather only what nature can afford to supply, don't compete with the foods wildlife need or that native people traditionally take, and harvest only fruits, leaves and other parts of plants that will regrow.

There are a host of good reasons to gather and eat foraged foods. Foraging establishes a connection with place, encourages restoration, supports local economies, reduces reliance on water and fossil fuels, protects wildlife habitat and sustains native species diversity, celebrates the region's unique food traditions, and contributes to health. Workshops and plant walks are also part of the project's efforts to educate consumers and would-be foragers. In the summer Patty and Teresa can be found at the Flagstaff Community Market, and they also want to publish a training manual and possibly a cookbook.

Because people are understandably challenged in preparing unfamiliar wild foods, each delivery comes with recipes that will work in any modern kitchen. Here's one for using eggs and *nopales*, the green pads of prickly-pear cactus: Singe the spines off one or two cactus pads. Slice into bite-size pieces, and sauté in butter. Stir up eight eggs and add a quarter-pound of shredded cheese and the sautéed cactus pieces. Scramble it all together in a skillet and salt and pepper as desired.

Each delivery basket brings a new surprise of "old friends" – plants of our neighborhoods that we often view as common weeds. These plants remind us that the value of farms and ranchlands can be measured not only in crops sown and livestock tended, but also in habitats suitable for native plants and animals.

SECTION FOUR
Sustainable Building and Energy

If everyone in the world aspired to the standard of living that residents enjoy in a town such as Flagstaff, Arizona, we would need the resources of four more planet Earths. What kind of houses can we build that everyone on Earth can inhabit without trashing the planet? There is no simple answer to this question, but this much we do know: building sustainably requires detailed ecological knowledge of natural systems, hardheaded assessments of political and economic realities, and some understanding of human aesthetic and ethical values.

Our standard of living is made possible by the availability of cheap fossil fuels. Fossil fuels are cheap partly because we purchase them without having to pay for the ecological damages resulting from their use. We don't have to pay much, yet, for the effects of global warming. We have deferred those costs into the future – ecological deficit spending.

Eventually, global oil production will peak and begin to decline and as oil gets more expensive, so will other types of energy. The availability of affordable energy in the future depends upon how well we conserve energy now. Conservation makes sense, economic as well as ecological.

Green building is a positive trend that focuses on easy-to-achieve goals such as energy efficiency, water recycling, and efficient use of building materials. Many good examples are showcased here. A project on the Navajo Nation uses abundant small-diameter ponderosa pine logs, timber that is generally valueless in most sawmills, to create much-needed traditional housing. But the road to sustainability is a footpath, not a super-highway.

Green building also embraces a variety of alternatives. Some are quite new, like the Earthships, which incorporate used tires. Others, like straw-bale building, have been rediscovered and brought up-to-date to current building codes. Perhaps most promising are new iterations of old technologies, such as the adobe building and restoration practiced by Mac Watson of Santa Fe. These types of green buildings often rely heavily on human labor. Consequently, many of these alternative homes cost more per square foot than conventional houses when built by contractors. This situation is bound to change, however, as the supply of cheap fossil fuels winds down.

Sustainability is approached even more powerfully, and realistically, in community. Witness Jack Ehrhardt's experience in rallying members of the Hualapai tribe to cooperatively build an Earthship. Such "barn-raisings" are big parties, opportunities for community-building – joyful labor. They point the way toward sustainability, which after all holds meaning only in practice, embracing all the ecological, social, economic, and spiritual realities of their place.

BUILDING FOR THE PEOPLE
Ed Dunn

Flagstaff, Arizona · By Ashley Rood and Peter Friederici

Ed Dunn

When Ed Dunn was a nine-year-old boy growing up in the then small city of Phoenix, his aunt came out from Kansas and rented an old adobe house for the summer. Ed still remembers how that traditional building was much cooler than the cinder-block house his own family lived in. It was beautiful, too. "I really loved going over there," he says.

Ed is still enraptured by houses, especially by houses built with the environment in mind. Designer, builder, educator, and visionary, Ed is now a tall, slender man who plays the guitar and bass on his time off, and is one of the leaders of the sustainable-building movement in northern Arizona.

What does sustainability mean to Ed? "It's about people," he says. "It's about improving the quality of life." Ed studied architecture at Arizona State University, but switched to a major in urban planning when he realized that most architects in Phoenix were designing nothing but strip malls and tract homes.

Ed became a teacher and taught inner-city students. He wrestled with the question: How do you communicate sustainability and tell students how they should live when they are afraid of going home – when they don't know where their next meal is coming from? "My students taught me that sustainability is about more than the environment," he explains. "It involves a myriad of questions about how to live."

Ed and his wife Teri moved to Flagstaff in the late 1980s. It was here he began to focus on creating spaces that truly worked for people. He began working as a handyman, and eventually began building houses. It wasn't long before he became interested in straw-bale houses. "What got me into it was the beauty of the thing," he says. "These homes have soul."

Not only did he like the way straw-bale houses looked, with their thick walls, earth-tone coatings, and hand-finished textures, there was also the fact that homeowners can really save energy with such homes. "Straw-bale houses built according to the principles of passive solar design," Ed states, "are three to four times more energy efficient than conventionally built homes." While these houses are not any cheaper to build, over the lifetime of the house the savings in energy is substantial – as is the savings in wear and tear on the Earth's resources. This long-term, holistic view of building materials is at the heart of a sustainable paradigm; it requires looking beneath the surface and beyond the moment.

Ed has come to love that sort of holism. "Many conventional homes," he says, "look like they don't fit in – they bear no real relation to their surroundings, or have an ill-thought out relationship." As an example, Ed cites the many houses east of Flagstaff that have stunning views of the San Francisco Peaks to the west – and that grow far too hot on summer days, and far too cold on winter nights, because of their huge windows. "More will be spent on such a home's energy in its life than on the cost to build it," Ed observes. His homes, instead, feature large windows to the south that trap solar energy, keeping the interiors warm at night. They still provide a scenic view of the mountains, just through smaller windows.

Each of Ed's straw-bale homes is unique. The straw structure and plaster finish allow for infinite creativity, from curvilinear walls to built-in custom benches. A recent project was an octagonal house east of the San Francisco Peaks that was modeled on a traditional Navajo hogan and designed according to Feng Shui principles. Its peaked metal roof echoes the diamond shape of the Peaks. Its walls are covered with straw-embedded earthen plaster that is beautiful and nontoxic. It is also low in so-called "embodied energy," or the combined "fossil fuel use" of a material over its lifetime, including extraction, transport, manufacture, installation, maintenance, and disposal. Inside, a cob wall built of dirt and straw curves around a bedroom; an indoor graywater system uses wastewater from the kitchen and washing machine to nourish plants. A highly efficient wood-pellet stove provides

Volunteers can always find something to do at the job site.

heat during infrequent periods of insufficient sunlight. It is a house that taxes the Earth's resources much less than a conventional house, and that very much belongs to the environment in which it is set.

Ed is still very much a teacher. He regularly offers workshops where participants learn the fine points of stacking straw bales or applying earth or cement plasters. He is a regular at local energy fairs and community gatherings and is a founding advisor of the Coconino County Sustainable Building Program. A few years ago Ed helped build the Willow Bend Environmental Education Center in Flagstaff, designed by local architect Paul Moore. Perched on the edge of a scenic canyon, this beautiful straw-bale building includes a south-facing wall covered in windows. In two places the windows are backed by dark "trombe" walls made of twelve-inch-thick concrete. The sun heats those walls during the day, and they release their heat all night long, maintaining consistent warm temperatures.

"It's amazing how constant the temperature is in here," says Center director Glo Edwards. "If we get two cloudy days in a row we might have to light a little fire in the woodstove, but that happens only about ten times a winter. It's so bright that we never have to turn the lights on." And the building is a teacher, too. Glo estimates that they probably get as many people dropping in to look at the building as to see the exhibits.

As a builder who cares about what happens to the Earth, to people, and to his community, Ed Dunn might well become dispirited by the direction of the building profession – build it fast, build it cheaply, don't worry about the long term. But he doesn't. He retains his steadfast involvement in the community and with the people whose homes he builds. Inspired by the creativity that exists in his profession and by the continual learning curve it offers him, Ed continues to strive for sustainability of the people, by the people, and for the people.

Straw-bale construction is eminently energy-efficient and community-friendly: it invites participation by volunteers.

MOVING FORWARD IN DIGNITY
Lilian Hill's Sustainable Housing for Indigenous People Project

Kykotsmovi, Arizona · By Rose Houk and Roberto Nutlouis

Young Hopi villagers learn an age-old method called cob building.

Three mesas in northeast Arizona have been home to the Hopi people for at least a thousand years. The grassy high desert may at first sight appear unlivable, but the land flourishes with life. Traditionally pueblo-dwelling farmers, the Hopi depend upon the land and the rain. Their sophisticated culture promotes balance and a humble life. They believe that to maintain balance – to live the Hopi way – they must constantly pray, keep their sacred covenant with the Creator, and caretake the land.

Inevitably, impacts from the modern outside world have created many challenges for age-old Hopi ideas and teachings. One young Hopi woman, Lilian Hill, is meeting these challenges head-on, with a vision to reinvigorate her people's older, self-sufficient lifestyle. She has chosen to start through one of the most basic human needs – housing.

Lilian, a member of the Tobacco Clan, lives in the village of Kykotsmovi at Second Mesa. She wears many hats – community organizer, college student, and poet – and her deep brown eyes are unwavering in her focus on sustainability and self-determination for her people. Living her entire life in the community, Lilian has seen firsthand the dire need for positive change. And so she decided to act. That action has taken the form of a project called Moving Forward in Dignity: Sustainable Housing for Indigenous People. The project's commitment, Lilian explains, is to create "affordable and culturally relevant sustainable housing for indigenous communities." It is designed "to address the critical housing need within these communities, whose members are often homeless or live in substandard housing."

The project was born of her family's personal experiences. Lilian's great-grandfather was a traditional builder, and her parents decided to obtain a "prototype" traditional home from the federal government's Department of Housing and Urban Development (HUD). Unfortunately, the dwelling suffers from leaking and cracked walls and roof, a malfunctioning electrical system, expensive central heating, thin frame walls, and bad concrete. "Such housing," Lilian asserts, "is of poor quality, ignorant of cultural needs, and expensive, particularly in relation to income and traditional lifestyle." On the Hopi Reservation, 30 percent of existing housing has serious deficiencies, she reports, and among the ten thousand Hopi living on the mesas, only 41 percent own their own home.

In general, houses on the reservation today include recently built cinderblock homes, manufactured trailers, and poorly built HUD homes. "The traditionally built stone and earth

The Sustainable Housing for Indigenous People Project provides housing – and a creative outlet – through the use of local materials and skills.

homes and traditional building techniques are becoming obsolete and are no longer being encouraged or supported," Lilian explains.

To address this situation, Lilian helped found the Sustainable Housing for Indigenous People project, with collaborators Paola Marcus-Carranza, the Cob Cottage Company, and Seven Generations Natural Builders in Bolinas, California. Lilian and her friend Paola Marcus took a course to learn a natural earth-building technique called the cob method. "Cob," she says, "is one of the world's most common construction materials." It's a "gourmet mix" of clay, sand, straw, and water, hand-sculpted while pliable. The term "cob" comes from Old English, and refers to a lump or rounded mass. Builders use hands and feet to form the material, which is then laid up without forms, in a process akin to sculpting clay. Near her home, Lilian found abundant clay and sand. Water comes from a pipe, but cob building, she says, is not water intensive. The method is especially well suited to northeast Arizona's climate because homes stay cool in summer and warm in winter. Also, it employs local materials and human skills, and results in lower maintenance and utility costs.

Lilian's innovative ideas combine traditional Hopi stone masonry techniques with cob building. In the home she and her family are building, the foundation and stem walls are of sandstone cut by her uncles at an old quarry at Hopi. The cob is laid on top of it. The north wall is straw bales with cob over them. Plastered earthen floors, niches in the walls, and a traditional roof of pine beams, branch latticework, sheep wool, and clay will complete the structure.

Lilian first spent a year of research on the building technique, then another year observing the cycle of the sun for best positioning of the house. She also took time to present the idea to the local community and was gratified by the "huge response, especially from the older women."

Then building began. Over time, some fifteen people have been actively involved in the construction, and many others have come simply to watch the process. "Our project," Lilian explains, "utilizes owner-builder and volunteer help from the local community, which will greatly decrease the total construction cost and will encourage unity and cooperation between the community and those involved." Some of these age-old methods have to be taught and re-taught. With help from Tim Reith and Sasha Rabin of Seven Generations, Lilian presents workshops to train people who are interested in this type of building. "I'm excited about these workshops," she says, "because they bring together people from different backgrounds and professions to exchange ideas. Everyone has valuable knowledge to share and positive change rests in such exchanges of ideas and understandings."

There's a bigger idea in all of this too. When people are involved in building their own homes, they receive a sense of place and appreciation that is greater than simply making a monetary investment in a place to live. "A home," Lilian observes, "is a sacred place where life evolves as families care for one another. Individual thoughts, prayers, and the teachings of the people evolve in a home." As people rediscover this important concept, Lilian is confident that they can regain self-sufficiency and depend less on the federal government.

"Ultimately," Lilian says, "we want to create a locally run organization to continue this work – one that provides people with training, the economic opportunity, and a long-term, effective structure in order to continue building sustainable and affordable homes in the future."

"I'm just one person, trying to find a solution," says Lilian modestly – a solution that fits with her community, the environment, and the Hopi way.

DESIGNING FROM THE OUTSIDE IN

Susie Harrington and w/Gaia Design

Moab, Utah · By Roger Clark

Susie Harrington and Kalen Jones

Susie Harrington sits cross-legged on the floor of her living room, which doubles as a yoga studio. "Buildings consume approximately one-half of the energy in this country," she says, then pauses, allowing her group of visitors to write notes. "One of my goals is to reduce that load by shrinking the size of buildings and the energy needed to control their temperature and lighting."

The Utah home of Susie and her partner, Kalen Jones, illustrates principles that guide their business, w/Gaia Design – self-described as "landscape architecture and sustainable building design, consulting, and educational services." For example, their home remains cool throughout Moab's notoriously hot summers, due to thick walls made from straw bales and a roof angled to shade the windows. That same roof allows the heat of the low winter sun to enter and be captured by the mass of adobe floors and earth-plaster walls. When a backup source of heat is needed during weeks of below-freezing temperatures, an efficient heat pump radiates warmth through the floor. Domestic hot water is provided by a solar heating system.

An enthusiastic teacher, Susie outlines other goals: "I try to avoid using building materials that require large amounts of energy to manufacture. Because the 'embodied energy' of concrete is quite high, I use it sparingly, only where it is needed in the footings, between the ground and the walls." She minimizes the amount of lumber required for framing by using post-and-beam construction. "Most of our windows extend upward to the base of the horizontal beams, reducing the need for additional lumber for window headers, and they are framed by load-bearing posts," she says. "Because lumber comes in standard lengths, I strive to make sure we aren't wasting it by needlessly cutting boards down to odd sizes."

Despite her home's many pragmatic virtues, it is its aesthetics that truly command attention. The adobe floor, with its light finish of linseed oil, has subtle undulations that feel delightful underfoot. The interior walls are sculpted of an earth plaster made with natural pigments that mimic the hues of Moab's slickrock landscape. Ancient juniper posts, salvaged from a nearby wildfire, accent the interior and support the ceiling, which is crafted from locally harvested aspen.

An inconspicuous loft serves both as a reading nook and as a guest bedroom. The design invites exterior elements into the home, making it feel larger than it really is. A thunderstorm forming over the La Sal Mountains to the east is framed perfectly in the living room's enormous circular window. The north-facing summer porch and kitchen windows overlook Moab's wetlands preserve, offering glimpses of passing waterfowl and the changing seasons. The soft sound of water dripping from a seep into a small but deep pool resonates through the southern window screens.

Susie began her career as a building designer, but soon realized a need to apply a more holistic philosophy in designing living space. "When outdoor parts of the home are given an equal or greater importance than the inside, a sense of connection to the outdoors results," she says. "Creating coherent outdoor spaces that interlock with the interior rooms encourages us to use these spaces."

Susie's buildings, like her yoga practice, are designed with serenity and balance in mind.

Her own garden includes an elegant alcove of intimate garden terraces. The walls of each level are expertly crafted from local sandstone, reminiscent of walls made by the area's early cliff dwellers. Ninety percent of the drip-irrigated garden plants are edible. The miniature pool at the base of the alcove collects rainwater from the roof and surrounding surfaces. Its shape, and the shade provided by wetland plants, are designed to minimize evaporation. Droplets from the pool's water are pumped into seeps that feed a hanging garden filled with columbines and meadow-rue. Hummingbirds and insects are everywhere. Another large circular window in the master bedroom looks out upon this idyllic scene.

Susie's clients include many new arrivals to the Moab area who are interested in custom-designed homes and landscapes. "One of my greatest challenges is to get them to think about building smaller homes," she says. "I encourage them to consider investing the savings that they can gain from reducing their square footage into better craftsmanship and landscaping." The result is usually small and superbly comfortable homes of 900 to 1,200 square feet.

Susie is generous in donating her time to local nonprofit groups. While teaching a four-week sustainable building course to a dozen college students from Vermont, the class and local high school students built a straw-bale greenhouse for Moab's Youth Garden Project. "It worked nicely with the college's curriculum, which places a strong emphasis on community service," Susie says.

Susie has been working on several other major projects. One is an environmental education center in Sonoma County, California. In addition to designing an education building that includes a cafeteria, offices, and an assembly hall, the project includes planning for a school garden, greenhouse, outdoor classroom, and a nature trail, as well as restoration of other parts of the site. Another project is a master plan for a hospital and senior living center in Moab that will integrate twenty-six acres of indoor and outdoor living space. Susie also hopes to increase her opportunities for teaching.

The scattered group of visitors meanders through garden pathways, enthralled and not anxious to leave. They look back onto the double-humped rooftop, artfully textured by shakes made from recycled plastic. Susie notes that the newly planted fruit orchard forming the top terrace is designed to obscure the view of a large tower holding high-voltage power lines. As in her practice of yoga, she seeks to balance opposites.

Susie smiles. "The whole point is to help people connect with their landscape, to make the outdoors more livable and to make people happy to be outside. I think people stay inside too much and that doesn't do anyone any good. In fact, it is a great malaise of our time." How firmly Susie Harrington believes that becomes clear in her answer to a final question about the size of her home. "Twelve hundred square feet," she replies, "but that's just the interior space."

WAY OFF THE GRID
Charlie Laurel and Margaret Spilker

Alpine Ranches, Arizona · By Ashley Rood and Rose Houk

The name "Earthship" might raise the question whether the alternative home design this word refers to is something of this world or perhaps the next adventure into space. In fact, these homes are quite earthbound and represent a self-sustaining, self-determined masterpiece of each owner's dreams. Charlie Laurel and his co-conspirator, Margaret Spilker, brought the Flagstaff, Arizona, area its first Earthship in 1996. The design, which incorporates rammed earth, used tires, recycled wood, and solar panels, is a truly sustainable home. With its curved walls, adobe plaster, and warm sunlight, the structure creates an overwhelming sense of calm and comfort the moment you enter.

Margaret and Charlie arrived at their mutual goals from divergent paths. Margaret came from a family of Nebraska farmers. She distinctly remembers her father attempting to convert to organic farming back in the 1950s, when such practices were almost heresy. It was this agricultural background that instilled Margaret's ethic of sustainable living. It derived from a farmer's need for common sense – figuring out what was really practical, what really worked in the long run. "Most important," she says, "is the firmly held belief that because things are done a certain way by most people does not mean that's always the best way."

This simple but defining ideology has freed Margaret to pursue new ideas. She lived in a progressive community in Santa Cruz, California, where she learned about solar energy and protested nuclear development. "Even more, I came to understand," she explains, "why we shouldn't continue to do what we are doing. There was information about positive alternatives, and this exposure to information and example was important to me."

Charlie Laurel

Meanwhile, Charlie spent his youth fishing, hunting, and catching snakes in the swamplands of southwest Florida. As he grew up he witnessed the heartbreaking destruction of his childhood playground by thoughtless development – millions of acres of pine forests and wetlands bulldozed, burned, drained, and paved. In high school he participated in an environmental education seminar in which he learned about the local ecosystem in the field and then helped preserve a threatened cypress slough as a regional park. Charlie's philosophy was also shaped by watching important adults in his life be driven by the demands of monetary goals. "I decided," he says, "that fulfillment through money didn't make sense to me. I had a head start on being skeptical of materialism."

When Charlie and Margaret met, they found they shared similar ideals that had arisen from very different experiences. During this time, in the 1980s, Charlie was building luxury homes in the San Francisco Bay area. As the two began to think about marriage and their future, buying an affordable home became a logical next step. Charlie corresponded with a friend who was building an adobe house, and that inspired him to research alternative buildings. He found a copy of *Mother Earth News* magazine. The cover – "Build a Home Out of Old Tires for $30 a Square Foot!" – caught his eye, and changed his path. The article was about architect Michael Reynolds' Earthship home design. Reynolds was the pioneer designer of the Earthship, an ecologically viable and affordable alternative home.

Charlie reexamined the work he was doing. "I was finishing up a 4,000-square-foot house, with seven split levels and nineteen-foot-tall glass walls that faced the wrong way, for two people to live in," he explains. "I thought about all the trees that went into that house, about all the energy that that house would need to stay in the comfort zone. I thought about people chaining themselves to redwood trees, global warming, and the increasing poverty of the Third World. The next day I was driving across the varied deserts of the Southwest to work on Earthships in Taos for a month."

That New Mexico experience inspired Margaret and Charlie to head to northern Arizona and build their own Earthship. As

Recycled bottles create a beautiful mosaic stained-glass effect.

Charlie explains, they were attracted by the region's sunny climate, and by the possibility of introducing the idea to a new area. "I was anxious to get one of these houses built as a showpiece," he says.

Though each Earthship is unique, most share fundamental qualities such as high thermal mass, an earth berm, passive solar features, and water collection and graywater recycling. Earthships are typically self-contained when it comes to water and energy, and they use locally available materials to attain thermal mass.

But Charlie and Margaret quickly realized that while the decision to build an Earthship was easy and obvious, other decisions were not so straightforward. The choices and compromises involved location, size, efficiency, materials, resources, and transportation. One of the most difficult decisions for them was finding the right location. Land in Flagstaff was expensive, but cheaper land outside of town would mean a longer commute and increased consumption of fossil fuels. After carefully considering the tradeoffs, they selected land twenty miles east of Flagstaff – a longer commute for Charlie, but shorter for Margaret's work in Winslow.

On their forty acres of high desert land they found abundant sunshine and plenty of space to live as they wished. Building codes in Coconino County did not prove an obstacle to their unorthodox construction. Some engineering matters had to be worked out, but mainly it was a matter of documentation and education. After satisfying the first requirements of design, engineering and permits, Charlie and Margaret launched into construction. The core materials for their house – 900 used tires – came from tire shops and a county transfer station. They hauled them to the site, packed them with dirt, and used them for the footings and load-bearing walls. As Charlie quips, the thousand-square-foot house is built of "steel-reinforced, rubber encased, rammed earth bricks."

The house is passive solar and has no mechanical heating system. There is a woodstove, but they didn't have to burn any wood the first five years, and have only used the stove a few times since. The south-facing walls are comprised of windows, the north side is earth-bermed, and the roof is heavily insulated. A small 400-watt photovoltaic system provides all the electricity needed to operate household appliances such as lights, stereo, computer, power tools, and washing machine. "We just don't use everything at once," says Charlie.

Rainfall in their location is a meager six to ten inches a year. Still, they manage to capture enough for drinking and other uses off the roof; the water is held in a 4,500-gallon cistern. In dry springs they sometimes have to haul water before summer rains arrive. Graywater is filtered through indoor planters; the plant roots oxygenate and purify the water, which is reused in the outdoor garden and in other locations.

The house contains beautiful details – used glass bottles set into an adobe wall send colored shafts of light from the bathroom into the bedroom; lavender, lemongrass, and avocado sprout bountifully from the indoor planters; a herringbone pattern of recycled wood elegantly graces the living room ceiling.

For people contemplating building a sustainable home, Charlie offers this advice: "Foremost, think small." It's not square footage, but quality that matters. Savings from building small can be invested in energy efficiency, solar or wind power, and alternative building methods. The amazing thing about their Earthship, Charlie says, "is that it is so simple." And in his mind it comes closest to his idea of true sustainability. "The future," he insists, "is not in high-tech substitutions, but in low-tech, human-powered construction."

Charlie and Margaret's Earthship home is an essential model of sustainable building, and it illustrates a way to put ideals into practice. In a world sometimes submerged in despair about the environment, these two people have shown that with thoughtfulness and fortitude, the positive is possible.

Reconnecting the People and the Forest
The Hogan Project

Cameron, Arizona · By Peter Friederici and Roberto Nutlouis

From most of the western Navajo Nation, in north-central Arizona, the diamond summits of the San Francisco Peaks are landmarks on the horizon. To the *Diné*, as the Navajo call themselves, the Peaks are one of the four sacred mountains that demarcate their traditional homeland. They're topped with snow for much of the year, and cloaked with a dark green forest.

During recent years of drought the snow has been off-kilter: it's come late and melted early. Many of the forests around the mountain, too, are out of balance. Once-open ponderosa pine forests have become dense with flammable vegetation due to more than a century of livestock grazing and fire suppression. This combination has worked together to eliminate the frequent but light ground fires that once kept the forests sunny and spacious. Now, when fire comes, it quickly climbs into the forest canopy and can char tens of thousands of acres. Since 1996 several large, devastating fires have sent towering plumes of smoke into the sky south of the small town of Cameron, about an hour north of Flagstaff.

In response to these ecological changes, the U.S. Forest Service and other land managers have launched an ambitious program of thinning the forests and reintroducing ground fires. But a major snag concerns the question of what to do with the trees that are cut, the vast majority of which are too small for conventional lumber mills.

In Cameron, one elegant solution to that problem is being shaped – a solution that is simultaneously addressing a grave need for affordable housing on the sprawling Navajo Nation. The tribal government recently estimated that 22,000 new housing units are needed on the reservation to replace flimsy mobile homes or old, decaying hogans – the round or eight-sided homes traditional to the Diné.

Why not, Mae Franklin mused a few years ago, use the small-diameter ponderosa pine logs from forest thinning projects to build modern, affordable hogans? Mae, who grew up in the Cameron area, works as a tribal liaison for the Forest Service and National Park Service. In 1999, she became one of the founders of Indigenous Community Enterprises (ICE), a nonprofit group whose mission is to work directly with indigenous communities to develop local economies that respect and incorporate traditional culture. One of ICE's first undertakings has been the "Hogan Project,"

Trees thinned from fire-prone pine forests can be a waste product – or a raw material.

which makes use of the abundant small-diameter timber to meet the housing needs of reservation residents. ICE worked with the Cameron community to develop a manufacturing plant that utilizes pine timber for hogan construction. The plant, SouthWest Tradition Log Homes, is a model of effective collaboration, as shares in the business are owned by ICE, the Navajo Nation, the Cameron Chapter, the plant's managing partners, and plant employees.

ICE continues to operate as a nonprofit with Mae on its board of directors. It finds affordable-housing funding for hogan

construction, especially for tribal elders living in substandard housing. ICE conducts financial literacy workshops that allow Navajo youth to learn basic financial skills and gain economic self-sufficiency through such new tools as mortgages on reservation land.

ICE also coordinates "community build" projects, in which high school students work on hogan projects to learn construction skills. Recently, for example, local youths helped build a hogan for Grandma Anna Cly, who had lived in the same hogan in southern Utah's Monument Valley for

forty-five years. The students learned everything from the ground up – from taking inventory of building materials to reading blueprints, preparing the land, framing, and putting up the walls. They watched how the plumbing and electrical work were done. Later, they were able to brief congressional offices on their participation.

SouthWest Tradition Log Homes now offers kits for octagonal hogans and more conventional log homes to buyers both on and off the reservation. The plant is run by Ron Taylor, who has spent more than twenty-five years working with log homes in the United States and Germany. Ron developed proprietary equipment that allows the company to use eight-inch-diameter pine logs for construction. That is the exact material that has the least value in traditional processing, he points out. "The smaller stuff can be used to construct pallets, and the larger material has value for dimensional lumber. We're taking the stuff in the middle that no one else really wants, and we're trying to make a value-added product out of it."

First the logs are milled perfectly round, like giant dowels, then dried in the dry desert air for up to eighteen months. They then receive a rounded, lengthwise "Swedish Cope" cut that allows one log to be laid atop another; finally, they are cut to length and saddle-notched to fit together. Leftover wood is cut for firewood, which is also in great demand on the reservation. It's a highly efficient process that produces little sawdust; most producers of log homes, in contrast, first square off round logs, then round them again, which wastes a great deal of wood.

The modular construction allows these hogans to be built quickly, which helps keep costs down – in fact, they are much cheaper than most log houses, and about the same cost as frame houses of equivalent size. The walls can be put up in as little as a day. But they'll last a long, long time. As Ron says, "There's a lot of warmth in log homes. And if your great-great-grandchildren come by, these structures will still be there. In that time a conventional home would have been replaced a few times, and a mobile home ten or twelve times. The value is there."

Every month the plant processes enough logs to build between five and eight hogans or cabins. But this rate of production is only a beginning. Ron hopes in a few years to be processing enough logs to build 175 homes a year, which he says might offer up to 145 jobs on the reservation, where jobs that pay well are scarce.

It's a compelling vision that connects the well-being of the Diné with that of the surrounding forests and that connects the past and the future. For centuries, native people here have relied on the region's forests and woodlands for firewood, fence posts, corrals, construction materials, plant foods and herbal medicines, and more. Converting the forest's excess into modern shelter is a powerful way of reaffirming the close link between a people and their place.

Left: A traditional Navajo hogan uses small-diameter pine logs from forest-thinning projects to provide much-needed housing.

GOING QUIET
Jack Ehrhardt

Kingman, Arizona · By Charlie Laurel

Jack Ehrhardt (right)

Jack Ehrhardt and his contracting company, ACE Builders, built the first "Earthship" in Arizona for a client in Dewey in the early 1990s. An Earthship is a self-sustaining, passive solar home made from used tires packed with dirt, cans, bottles, and other discards. The home was designed by architect Michael Reynolds of Taos, New Mexico, and features solar electric power, rainwater roof collection, and indoor graywater garden planters that filter wastewater while growing food and flowers.

Then Jack built his own Earthship home in the Cerbat Mountains above Kingman. It just made sense – environmental, economic, and common.

Jack Ehrhardt values independence. Earthships offer freedom from utility bills and mortgages, but more than that, they reflect Jack's independent mindset. "I don't participate in the Euro-American rituals – there is no Santa Claus and the Easter bunny doesn't lay eggs," he says. "That's part of being able to see clearly. I don't think on cue and that allows me to be

pretty free." Jack is a big man in all dimensions – huge in body and heart, with a big laugh and a readiness to take on big challenges. It's hard not to liken his massive, yet gentle persistence to that of an ox, but his quick wit and intelligence resist such comparisons.

Several years ago Jack drove into the town of Peach Springs, Arizona, for the first time. He found the administration building for the Hualapai Tribe and asked, "Are you guys interested in energy-efficient, sustainable building?" They said "yes," and then the council got together and three hours later Jack gave a talk. "Then they told their natural resources department to find a grant to get one of these built," he explains. An Environmental Protection Agency "Jobs through Recycling" grant eventually got the project underway.

With the Hualapai, Jack faced the kinds of problems typical of rural areas with high unemployment and scant resources for training programs. Sometimes workers failed to show up, and the project was vandalized more than once. "It was difficult, but it was good. The experience was all beneficial – school kids came down and worked on it, they did the can walls, the bottle walls, they learned about recycling, they got to do the earth plasters. We literally got the tires from the community. Everyone was gathering tires from ravines and people's yards, wood from dismantled buildings, windows from military reutilization. It was really a good time; it was really a good feeling. It seemed to me to be one of the best times things felt around here."

The 1,200-square-foot Hualapai Earthship now functions as a tribal office space with seven solar-powered workstations. Jack now serves as the tribe's Planning and Economic Development Director, and is working on developing renewable energy projects, such as a wind farm, to generate revenue and create jobs on the reservation. In 1999, Jack pulled together one of the most unusual collaborations in the field of sustainable building. He brought together the rebel architect and counterculture icon Michael Reynolds with the top brass of the Arizona Army National Guard. Colonel Doug Brown was committed to getting a 5,000-square-foot office building constructed using recycled materials – tires in particular – for the Guard's base in Phoenix. It didn't seem to matter that the right people for getting it done were the

Made from things most people throw away, Earthships embody sustainability.

long-haired Reynolds and Ehrhardt, even after Jack told them about his conscientious objector status during the Vietnam War.

Jack laughs when recounting the story. "At that point they said 'we have no labor, but we have some money for materials, and a very limited budget. And, by the way, your labor force will be Sheriff Joe Arpaio's drug-infested, dysfunctional prisoners.' I said, 'Oh, bring it on! I'm the man for the job. Give me the most impossible task.'" Jack rolls his eyes and laughs: "What a circus. I had to cut the locks off the fuel depot to get fuel for the tractor, and they tried to get me arrested. I said, 'I don't care what you guys do, I'm gonna get this building built!' Fortunately the colonels stood behind me, and through a long, long process we got the building built."

As for the "drug-infested, dysfunctional" labor force, he says, "Every single prisoner learned things that they never dreamed were possible. Some of them never even worked before, they'd simply been dealing drugs, and they said 'Wow, so you can build things like this and you can use solar energy.' And I take a big fire hose and shoot it across the sky and it makes a prism, and I say, 'Look guys, what you're breathing.' And if you can picture thirty inmates, in Arpaio's uniforms, with their mouths open, going, 'So we're breathing in rainbow energy!' I said, 'Yes, men!'"

Jack's building projects all become educational forums – opportunities to weave community values around sustainability. The Ehrhardt's Earthship home in Kingman has served as the hub of a youth education summer camp focused on teaching kids about renewable energy and conservation.

Jack doesn't hesitate to get involved with local issues, such as organizing a successful campaign to prevent the construction of a toxic waste incinerator. He has also served on the local planning and zoning commission.

"If we speak from the heart, so the people sitting at the desks can feel it, then they make the right decisions," he says. "That's what activism is about – making life exciting and participating. Life is so much clearer and vibrant when you do that."

Even as a family man responsible for raising two children, Jack didn't feel the need to compromise his values for the sake of secure employment in conventional construction. But he doesn't consider himself particularly courageous. "I don't know if it's a life purpose or just going calm and paying attention to something as simple as what the church was saying, and your parents taught you: to do good. And then you become an adult and throw fifty percent of it away and compromise it. I'm doing what I was taught. It's no big deal. It doesn't make sense not to do what we're doing: seeking peace and doing good. I don't know why other humans don't feel it, or why they don't choose to go quiet and contemplate and sense their connection to the natural environment and feel the responsibility. It's fun to give your life choices a priority to where they make a difference toward doing something about the whole family of planet Earth and the whole cosmos that we live in."

And with that philosophy, one man keeps his life, and his Earthships, on course.

DIRT RICH
Mac Watson

Santa Fe, New Mexico · By Charlie Laurel

The story of adobe building is as old as dirt – as old as sun-baked mud bricks, and almost as widespread as human beings themselves. "Eighty to ninety percent of the structures in the world that people live in are earthen structures. It's only in the Western, developed countries that we've forgotten how to use this material," Mac Watson says.

Mac has spent thirty years restoring old adobe buildings and teaching people traditional adobe building techniques. He wonders how we could, after thousands of years of use, lose the knowledge of this most basic and universally practiced building skill within only two generations – and how, in his hometown of Santa Fe, New Mexico, only the rich can afford homes made from mud bricks, while the poor inhabit the products of industry – mobile homes.

Mac's own adobe home tells much of the story: "The house was built by local people who led an agrarian lifestyle, so it was a tiny beginning of a farmhouse in 1910 in the upper Santa Fe River canyon – two rooms and a kitchen." In those days farmers still knew how to make adobes and how to build with them. They built their own homes using simple hand tools. They were subsistence farmers growing and trading diverse crops. Their homes and their livelihoods were products of the earth. "That's the relationship between culture and

agriculture," says Mac. "If you have a culture that is agriculturally dependent and is performing sustainable agriculture then it's very likely that they'll be living in sustainable structures as well."

Mac Watson

For Mac, adobe is the ideal sustainable building material "because it's there. It's hard to say that it's an unlimited resource," he laughs, "but there's still plenty of dirt! And there's not a lot of petrochemicals and steel and the other things we use to build our structures." Still, adobe is more expensive to build with than stick framing because it is more labor-intensive. Stick framing and other industrial building methods are cheaper because of the availability of cheap fossil fuels. Fossil fuels allow machines to do work instead of people, but they also have severe environmental impacts. Sustainable building practices are therefore, almost by definition, more labor intensive and more expensive, but environmentally less costly. "People now are not in a position where they can stop working to build their house," he

In the Southwest, adobe is often the ultimate local building material.

explains. "The only thing that poor people can get financing for is a mobile home where the interest rates are outrageous and the building is falling apart by the time you pay it off. But you can get into one quick. They make it real easy." But no one makes it easy to fix up an old adobe house. He explains, "Banks won't lend you money with an old adobe as collateral. They want something they can repossess. So, unfortunately, we're seeing more and more adobes disappearing from the landscape because of the economic conditions. When you lose the agricultural base, then you also lose the architectural expression of that agricultural base."

Against this backdrop, Mac found that he wasn't interested in building new adobe homes for rich people "who don't really need a third or fourth home." He opted instead to work on preserving and restoring historic adobe buildings. He worked for ten years as a volunteer for the nonprofit Cornerstones Community Partnerships and then took a staff position with the organization. The mission of Cornerstones is to help communities restore their historic buildings, mainly historic churches.

"Part of the idea of restoring the churches was to reestablish community values," Mac explains. "The community would ask me for help in restoring their church and it was obvious that the only way that it could be done was to gather the forces of the community in order to do the work." He continues, "Community values for me are primarily cooperation and mutual respect and a willingness to work together to accomplish a community goal. Doing restoration work on an adobe structure is a perfect, perfect place to develop or redevelop community values because you get people working together, and they drop their age-old grudges and find out they can get along with each other. Sometimes that spreads to other kinds of community efforts."

Mac cites as an example his work on a Penitente place of worship in rural northern New Mexico. Penitente brotherhoods came into prominence in New Mexico around the beginning of the nineteenth century, when there were few Catholic priests in the region. They built *moradas* – windowless adobe structures for men's religious rituals. Mac assisted with the restoration of a unique Penitente building constructed by pouring adobe mud into wooden slip forms along with cobblestones. The whole community participated in the work, and a renewed interest in Penitente religious practices emerged when the building was done. Some Penitentes from a neighboring community came and gave instruction in the rituals, and many young men of the community took up the old religion.

"It's one of those cases where architecture had its impact on culture, so we had a revival of traditional architecture techniques and then a revival of community traditions," Mac says. "Northern New Mexico has some of the highest poverty rates in the country. There are tremendous social problems, huge drug problems, crime problems. Getting the guys involved in the morada distracts them from going out and getting involved with drugs and gang warfare."

Mac also sees an important role for restoration in ecological sustainability. "Take care of what you have," he advises. "Save what you have and preserve it, and you're not eating up all your resources by throwing stuff away. Most historic structures are within an urban landscape. Usually when a historic structure is torn down, what replaces it is a parking lot. It's almost inevitable. You're going to have an asphalt parking lot for years and years and years."

Restoration is a very conservative enterprise, changing as little as possible of the original structure under repair. For Mac, this means saving the original doors and windows whenever possible. "Doors and windows are some of the most significant historic features in a building," he explains. In his garage workshop Mac illustrates his thinking, showing an old wood-frame window sash. "Windows are the first thing that everyone wants to throw away. They want to replace it with a new 'energy-efficient' window. But we're talking here about long-term economy… the energy-efficient windows will never pay for themselves in savings."

Mac Watson hopes to renew people's appreciation of craftsmanship and tradition while restoring old adobe buildings. He believes that old buildings give continuity to our lives, connecting us with generations past. And caring for old buildings, in turn, connects us with generations to come.

SOMETHING NEW – AND OLD – UNDER THE SUN
Hopi's NativeSUN

Kykotsmovi, Arizona · By Peter Friederici

Technicians install solar electric panels.

During most daylight hours the sun shines strong on the Hopi mesas of northeastern Arizona, ripening the corn, warming the houses, and forming the clouds that nourish crops. For the last twenty years it has also powered the electrical systems of a growing number of homes, thanks to a locally based business that's pioneered the practice of bringing solar electricity to remote locations.

NativeSUN was founded in 1985 as an offshoot of the nonprofit Hopi Foundation, and became an independent for-profit business in 2001. It now has two offices in Arizona and one in Colorado, and has installed about 450 residential power systems on the Hopi and Navajo reservations, as well as larger systems for such institutions as the Tucson Unified School District and the Flagstaff office of the Grand Canyon Trust, a regional environmental group. The project's success has not only allowed hundreds of customers to acquire energy that is clean, nonpolluting, and reliable; it

has also become a good source of employment on reservations where high-paying jobs are scarce. During busy periods, the company employs up to eight solar technicians who install new electrical systems and service existing ones.

The project owes its inception to two factors. The Hopi and Navajo reservations are remote, and many residents have been bypassed by traditional electrical providers. On Navajo tribal lands, in particular, homes are widely dispersed, making it difficult and expensive to hook up to the electrical grid. In addition, several Hopi villages have explicitly forbidden the installation of transmission lines. These traditional people do not want to lose land by providing rights-of-way to utility companies, and many believe that the electromagnetic fields emanating from the lines would disrupt the atmosphere and balance of the villages' ceremonial plazas – to say nothing of the aesthetic impact of the lines. As a result, more than ten thousand residences on both reservations remain off the electrical grid. In many cases residents still run lights and appliances; they simply power them with noisy, gas-powered generators.

Right: Solar panels can bring electricity to areas where connecting to electric transmission lines is difficult, expensive, or unpopular.

SOMETHING NEW – AND OLD – UNDER THE SUN

In this sunny, wide-open setting, solar energy is an ideal solution. NativeSUN has specialized in installing small-scale systems that allow homeowners to live independently of the grid. "I don't have to pay for electricity," one client has said. "I recommend it to other people. Once you pay for the system, it's yours." Most of the systems cost $10,000 or less – still a considerable investment in this job-poor area. NativeSUN assists clients in paying for equipment and installation with its own revolving loan fund and by helping to locate other sources of credit. The freedom from monthly utility bills is highly compatible with Hopi traditions of self-reliance.

A new energy technology meets ancient building methods in a venerable Hopi village.

about energy use, as well as such energy-saving tools as compact fluorescent light bulbs.

Doran is from the village of Hotevilla, on the Hopi reservation, and he helped start the Hopi Foundation after attending college and studying electrical engineering. He is consistently gratified by the response of residents who discover the joys of having their own electrical system.

"Individuals who have never had electricity love it," he says. "When we install it and flip that switch, you should see their faces – they're so proud and happy. I think that's what keeps us in it, is that excitement." The pride is especially pronounced when considered in the light of history. In many cases, residents remember past promises made by federal agencies that electricity would arrive within a few years. Often it did not. And so they are gratified to discover that even on the ancient Hopi lands there is something new under the sun – albeit something very much in keeping with venerable traditions.

Solar energy is efficient, but it can take a bit of getting used to in a home. People who have lived in towns on the grid, where turning on lights and appliances is as simple as flipping a switch, must learn the mindfulness needed to live happily with a smaller source of energy. "The initial feedback we get from individuals who have lived in the city can be that the system doesn't produce enough electricity," says Doran Dalton, NativeSUN's director. "They try to run too many appliances and then run out of electricity. It's the law of natural consequences. But after a couple of months, we have no more complaints. They learn: if you're not in a room you turn off the light. And if they really need more energy, they get everyone to pitch in and add on to the system with more panels." NativeSUN technicians do assist their customers after installation by offering education

"One woman had her house wired for electricity once – when her village was told that electricity would be coming within three years," Doran Dalton says. "When we were finished with the installation job she said, 'After thirty-five years I can't believe there's finally electricity going through those wires.'"

DOING GOOD FOR THE WORLD
Southwest Windpower

Flagstaff, Arizona · By Rose Houk

Andy Kruse's story is a classic one: he and a friend started messing around with wind turbines in a garage "out in the cinder hills" near Flagstaff, Arizona. The year was 1987, and he and David Calley had visions of harnessing this source of renewable energy for the betterment of humankind. David was making wind generators even when he was in high school in the early 1980s. With his technical expertise, and Andy's marketing background, the pair founded Southwest Windpower and made a small wind turbine that they called the Windseeker.

Seventeen years later, David remains company president and Andy is vice-president. They bought an old bowling alley in Flagstaff and converted it into a manufacturing plant where some fifty people are employed making four basic wind systems. Each new version has involved close evaluation and improvements through constant research and development.

The company's modest offices are near the plant. On the shelves in Andy's office are books with titles like *The Next Great Thing* and *State of the World*. Maps on the wall show average windpower estimates in Egypt and the most favorable wind areas in the Dominican Republic. Though the month is March, Andy's calendar remains on January.

Andy is like a constant-motion wind machine himself. He takes a moment to check something on his laptop – a link to anthropologist Richard Leakey, whose home in Africa is equipped with a Southwest Windpower turbine. It is such small-scale, individual-sized devices that the company now produces at its factory.

What motivated them to start the business? "I saw an opportunity to sell this around the world," Andy replies. Two billion people are without power, he adds, and energy is needed in remote areas. "It's about providing power, based on renewable energy, in small amounts," he explains. "We're in the wind business, but we're really in the electrical business." The fundamental importance of having electricity means people will have things like water, radio, telephone, and television. Most of these are beneficial to a way of life, but Andy also fully recognizes the cultural implications.

Southwest Windpower turbines have popped up in all corners of the globe, energizing research facilities in Antarctica,

Andy Kruse and David Calley

SUSTAINABLE BUILDING AND ENERGY

pumping water for nomadic herdsmen in Mongolia, and electrifying streetlights in Tokyo. Japan, he says, is one of their best customers because of the country's commitment to renewable energy. The sleek, three-bladed turbines even adorn the tallest skyscrapers in the city. Andy turns to his laptop again and clicks on color photos of the company's machines attached to recreational vehicles, sailboats, and even a friend's tree house. They also see applications in other hard to reach or extreme places such as mountaintops and offshore platforms.

Obviously a busy man, Andy has a boyish ease about him, whether it's talking to government officials about local ordinances, speaking with state commissions and big utility companies, or dealing with World Bank executives. And though the world is his marketplace, he vows that the "first vision" he shares with David is to look into rural areas in the United States, especially Indian reservations, which in some places still lack electricity. Two hundred hybrid systems using wind and solar photovoltaic cells have been installed across the Navajo Nation in northern Arizona and New Mexico. The second and bigger market is what he calls "grid-tied." With the introduction of a new two-kilowatt turbine, an entire house can be powered, and the machine will pay for itself in five years. When the wind is not blowing, people can buy power from the electrical grid. When the wind is blowing, the meter runs in reverse back into the grid.

Andy is the first to admit that wind power is not for everybody. Because it's not available on demand, wind will always be a secondary source of energy. "You won't see wind turbines on every rooftop," he allows. But he follows with a quick recitation of a statistic: "there are seventeen million homes in the United States on an acre or more of land." That's what Andy sees as a prime market, and he'd like nothing better than to capture a large portion of it. The secret to success in the United States, he believes, is to make wind power convenient. The technology must be reliable and cost-effective for people to choose to invest in it.

The basic concept behind harnessing the wind hasn't changed much through the centuries, but the technology has been greatly transformed. Wooden blades have given way to composite materials, and alternators rather than generators are used, making the turbines highly energy efficient. The one "weak link," explains Andy, remains storage in batteries. Though other possibilities have been explored, the lead-acid battery is still the cheapest and most reliable. But many of them are needed, and they have to be replaced periodically.

Before recommending wind turbines, Andy makes sure people first think about conservation. He wants people to understand how they use electricity every day, so that if and when they do commit to wind power they'll already be using the most efficient appliances.

Andy admits that he once thought of the wind as a nuisance. But since he's gotten into the business of supplying small-scale wind power, his feelings have changed toward this powerful force. Now, he says with a smile, "When I see a wind turbine spinning, I see someone doing something good."

Left: A Southwest Windpower turbine is ideal for powering remote sites, whether on or off the electric grid.

SECTION FIVE
Regional Food Heritage

Wild and cultivated organic foods are a true pleasure, and not just because they are healthy to body and environment. Whether travelers or local residents, people on the Colorado Plateau can now find a number of places like those showcased in this section where they can sample the foods indigenous to this region in style and in good taste.

One such place stands in Winslow, Arizona, in the historic La Posada Hotel. Here, at a restaurant called the Turquoise Room, Chef John Sharpe works magic with authentic foods of the region. He has made connections with many of the residents surrounding him – Hopi, Hispanic, and Anglo – from whom he buys corn, cheese, and meat to transform into delicious, intriguing menu selections.

Sixty miles west of Winslow is the Flagstaff Community Market – a weekly gathering of regional growers who set out their vegetables, fruits, nuts, herbs, tortillas, and tamales for sale. It's a friendly, easy-going event, set to the tune of trains rumbling by the downtown parking lot. One of the tables at the market is laden with baskets of apples from Garland's Orchard. Garland's trees grow in the confines of the cliffs of Oak Creek, south of Flagstaff. While the orchard is eighty years old, it has been cultivated solely with organic means for the past dozen years due to the committed efforts of Rob Lautze and Mario Valeruz. The rewards of their hard work can be seen, and tasted, in the wonderful apples.

Young's Farm, down near Prescott, Arizona, is another historic farm whose products are beloved among loyal customers. Four generations of Youngs have been farming in Lonesome Valley, selling their yields directly on the property at the roadside stand, farm store, and restaurant. The Youngs have survived as a family farm even amid increasing pressure from development in one of the fastest-growing parts of Arizona.

In another Four Corners state, Utah, a sign along the road into Moab reads, "Ye Ol' Geezer Meat Shop." Here, Rich and Pat Evans run an old-fashioned butcher shop that offers locally grown, naturally raised beef and pork. Rich butchers and dry-ages the beef cuts, and grinds hamburger fresh each day. Rich feels great pride in what he does and his customers can taste the difference. Like the other celebrants of local food culture you will read about here, his products are firmly rooted in a particular place.

CELEBRATING SOUTHWEST FLAVORS
The Turquoise Room at La Posada

Winslow, Arizona · By Gary Paul Nabhan

John Sharpe

When you leave the noonday heat of the Painted Desert and come into the cool, colorful halls of the historic La Posada Hotel in Winslow, Arizona, it sometimes feels as though you are going back in time. Designed by pioneering Southwest architect Mary Colter at the end of the 1920s, La Posada was once one of the renowned Harvey Houses along the Santa Fe railroad, and is now recognized as a National Historic Landmark. But beyond visual beauty, La Posada offers exposure to a variety of flavors and fragrances that are uniquely and authentically southwestern. From the heirloom gardens and quince orchards surrounding the hotel to the tastes of the Turquoise Room restaurant inside, La Posada is helping revitalize and celebrate the region's finest food traditions.

John and Patricia Sharpe's Turquoise Room was named after an historic dining car on the Santa Fe's Super Chief, the train that brought everyone from Albert Einstein to Hopalong Cassidy to Winslow in the 1930s. The ever-changing menu at the Turquoise Room has been widely praised in venues ranging from NBC's *Travel Café* to Slow Food USA's *Snail* newsletter to *Gourmet* magazine. Chef John Sharpe, who spent a quarter-century bringing nouvelle cuisine and New Southwestern Style to restaurants in southern California, left metro L.A. behind in the fall of 1999 to come to Winslow. His big-city colleagues laughed at the idea that he would find a sizeable constituency for his culinary innovations in Winslow. Since he opened the restaurant in 2000, though, John has been dishing out about a thousand meals a week. A number of his fans regularly drive 100 to 180 miles roundtrip just to partake of his latest experiment with regional flavors.

But it is not merely the volume of business nor the distances traveled by customers that serve as the best indicator of the Turquoise Room's success; rather, it is the unique partnerships that John and Patricia have forged with Hopi elders, Hispanic ranchers, Anglo goat-cheese makers, and wild foragers in the region. The Turquoise Room is perhaps the only restaurant in America that regularly offers a suite of ancient foods made from native crops, from blue paper-thin piki cornbread and white tepary bean dip to roasted Hopi sweet corn. Wild foods such as native greens, mushrooms, elk, and bison regularly roam the kitchen and the menu of the Turquoise Room.

A restored architectural classic, La Posada houses a culinary gem, too.

John Sharpe does not simply develop recipes for the restaurant's entrees: he also grows some of its foodstuffs at home, and personally selects other ingredients from the offerings of farmers gathered each summer at the Flagstaff Community Market. "My entire childhood following World War Two was spent gardening," he recalls of his early life in the English countryside. "That experience was essential to shaping how I select vegetables and fruits to use in my kitchen."

In addition to his own food production at his home on the edge of Winslow, John has guided his neighbors in the growing of Hopi sweet corn, and his friends at Stargate Valley Farms south of Holbrook in their herbal seasoning of goat cheeses. His feedback to farmers and ranchers – and his eager use of their pilot products – has provided needed impetus for advancing sustainable food production on the Colorado Plateau.

It is a mouth-watering experience merely to read the Turquoise Room's ever-changing menu. When his late-autumn menu includes a chutney that draws upon quinces that Mary Colter planted on the grounds of La Posada, the rich caramel flavor echoes the chemistry of the very soil that the restaurant sits upon. When John and his sous chefs prepare *chile en nogadas* – stuffed ancho peppers in a creamy nut sauce topped with pomegranate seeds – it is easy to see why this dish was featured as an aphrodisiac in the Mexican film *Like Water for Chocolate*. And when summer comes around, the Flagstaff Community Market salad becomes a feature that changes in its elements each week, but always offers its partakers an opportunity to savor greens, tomatoes, and other fruits as they are bursting with ripeness – a sensibility that cannot be gained by eating a salad containing tomatoes that were shipped green halfway across the continent, then gassed with hormones to turn them red a few days before they reach a restaurant.

Unlike many other chefs interested in Native American and heirloom crops for use in restaurants, John has a taste for truly authentic traditional foods unadulterated by fads in the marketplace. His support of Hopi tribal members and their use of a traditional piki stone for making blue corn wafer breads is exemplary of his approach. Even before he relocated in Arizona, John was pioneering "Foods of the Americas" native feasts at restaurants and museums.

"To find a hardworking chef is easy, to find a passionate chef is refreshing, to find a chef that cares for his community, his staff, the history of the region, and the integrity of local foods is highly unusual," says Paul Buchanan of the American Food and Wine Institute, who once worked with John in California. "We should all be so lucky to have a treasure like Chef John Sharpe cooking in our neighborhood."

Should anyone scoff that such concerns and interests are too esoteric for the average American restaurant-goer to understand, all they need to do is eavesdrop on the conversations around Turquoise Room tables. John's self-composed menu notes educate his visitors on the origins, history, and diversity of southwestern foods.

He has not at all regretted his move from the urban intensity of southern California to the rural life of the Painted Desert: "It was staggering to learn that I had landed in an area that has five million visitors a year, most of whom travel Interstate 40, and there was hardly any place of quality for them to eat," he says. It was once considered a culinary wasteland by food writers and restaurant critics; now John Sharpe has put the Painted Desert back on the map.

MIXING ART AND FOOD
Cow Canyon Trading Post and Restaurant

Bluff, Utah · By Roger Clark

"I came to Bluff eighteen years ago and fell in love with this place," reflects Liza Doran. "We've slowly built our trading post and restaurant business by working with people in our community and their families. These gray hairs show that it hasn't been all that easy. But the wonderful relationships we've developed here, and its incredible beauty, have made it all worthwhile."

Bluff, Utah, is a long way from Liza's previous home in Ghent, Belgium. Set amid the red sandstone bluffs that line the San Juan River in far southeast Utah, the town supports a lively community of about 250 people. About the same number of hardy Mormon pioneers founded the town in 1880, having pioneered an arduous wagon route that traversed the steep canyon of the Colorado River. Most of their descendants have long since left town, replaced by an eclectic mix of river runners, artists, archaeologists, authors, retirees, small-business owners, and one dobro-playing hay farmer who recently planted a vineyard.

Liza's love affair with redrock country began in Ghent, where for five years she sold Zuni jewelry "a la tupperwear parties." Jim Ostler, her business partner, managed the tribal store in Zuni, New Mexico. Along with jewelry shipments, he sent her pictures of Bluff, a place she had never been. "He said that there was an old trading post that he was interested in buying," she recalls. "When I returned, we looked things over in one afternoon and made the decision over a beer in the Nada Bar, which was called the Silver Dollar back then."

They bought the trading post from Rusty Musselman, who had built it shortly after World War Two. It is a single-story building made from cut sandstone that was salvaged from structures left by the original Mormon settlers.

Liza Doran

The couple opened the trading post and featured some work of Zuni artists, but they soon began to focus on showing and selling the works of the many artists who live in the neighboring Navajo Nation. A year later, Liza opened a restaurant in rooms to the rear of the sales area. Several of the smaller rooms serve as galleries for displaying both fine art and shelves of books for sale, including rare classics about the Four Corners region. The main dining area is an enclosed porch that looks out over the historic Jones Farm,

Cow Canyon's menu includes dishes made from fresh local and regional ingredients, artfully served.

newly protected by a conservation easement. Liza and a determined group of Bluff residents raised nearly a million dollars to prevent its development. "The farm defined the town and gave it a sense of place," she says. "We couldn't stand the thought of losing it to developers, piece by piece."

From the beginning, Liza has employed cooks and servers from two families of artists. "I'm sort of the orchestrator and coach in the kitchen, but everyone pitches in. Last night Ruby [Warren] made lamb stew, and then Daisy [Buck] came in and decided it needed some chile. We rarely make breads the same way twice. It depends on what we have on hand. We joke that we are all *nat'aanii* in the kitchen. That's the Navajo word for 'boss.'" When Daisy appeared to help prepare for

this season's reopening, Liza's son Elliot greeted her with "*nat'aanii*." He loves to sing traditional Navajo songs that he has learned from Ruby and Daisy since childhood.

Liza's simple, handwritten menus always offer a vegetarian and a meat entree and homemade soup and bread. She also features seasonally available greens and vegetables from her garden, traditional lamb stew, and other regional specialties. She grows herbs, tomatoes, eggplants, and peppers and is sometimes able to buy vegetables from other local growers. "I have to drive three hours to get romaine lettuce, so if I don't have an ingredient, I just deal with it," she says. "I'll go out in the yard and work with what's there. It's like Ruby: she can make spanakopita or overhaul your truck. You learn to be resourceful around here."

The artwork Liza and Jim show shares the same local focus. "Our first Navajo rugs came from Jim's collection from when he worked at Hubbell Trading Post and a one-time purchase from the auction at Crown Point, New Mexico," Liza says. "After that, the word got out, and local weavers have kept us in constant supply ever since." Today the trading post is filled with unique creations that have been acquired through Jim and Liza's long-term relationships with about twenty-five artist families who live in the area. Their inventory includes beautiful Navajo rugs woven by Mary Jim, pitch-glazed pottery made by Rose Williams and her family, Thomas Begay's wooden sand paintings, and Gregory Holiday's exquisite ceramics, with designs inspired by the pottery

Right: Cow Canyon Trading Post and Restaurant has forged long-lasting relationships with nearby families.

MIXING ART AND FOOD

the region's ancient cliff dwellers left behind. "We're lucky to be living near a lot of very talented and imaginative people," Liza says. "Most of our art comes from within a fifty-mile radius."

Liza enjoys a special relationship with several families of folk artists. "Delphine Warren's work is hilarious," she says. "She paints carved sandstone figurines of Navajo bureaucrats, bathing beauties, tourists with cameras, George Bush, and Mormon missionaries. Her father Homer, who's gone now, started carving horses out of naturally spotted sandstone. Then her mother Irene would paint saddle blankets on them.

Jim Ostler with pieces from local artists.

The trading post is open year-round, and the restaurant is open from April 1 through October 31, serving dinner several evenings a week.

Like the quirky folk art inside, Cow Canyon Trading Post and Restaurant alerts those who arrive from the north that this community might be a bit different. Parked permanently in front of the trading post is a 1949, faded green Dodge pickup truck. On its side is printed:

King Cole Farms
Home of King Rastus
Registered Quarter Horses
Littleton, Colorado

"Hilda Bitsillie was one of the first to bring me mud toys. She has a remarkable face, and when she unpacked her toys you could see her delight in them," remembers Liza. "Marie Oliver is always bringing in funky things like figurines with eyes painted by stabbing the brush straight in to create star-like eyes. When we talk, her grandkids translate in this crazy, homegrown kind of sign language. I once got her to laugh when we were talking about a man's figure that kept tipping over. The woman's figure that went with it stood up because she wore a wide skirt. I made a hand sign to suggest that he could stand up by leaning on her."

"It belonged to a man named Ralph who settled here in the 1950s," Liza says. "He was a really beloved fellow here in town – a geologist and really talented musician, back in the days when Bluff was a bit more raggedy than it is now. He headlined Uncle Ralph and the Goatheads."

Proud of their local heritage and determined to sustain personal relationships that make their community unique, Liza and Jim know that they live somewhere special – and they aim to keep it that way.

COMMUNITY CONSERVATION
Young's Farm

Dewey, Arizona · By Ashley Rood and Rose Houk

In an era marked by unfettered land development and tough economics, family farming is quickly becoming an endangered way of life. All over the country fertile topsoil is rapidly giving way to concrete and asphalt. The operators of Young's Farm in north-central Arizona know this better than most.

Young's Farm has been held by the same family for more than sixty years, through four generations. The farm is unusual not only for its staying power, but also because it has become the community center of the small crossroads town of Dewey, south of Prescott. But this longevity is in jeopardy with the encroaching development of Yavapai County, one of the fastest-growing parts of the state.

The history goes back to the mid-twentieth century, when newlywed Elmer Young, just back from World War Two, moved to Lonesome Valley in 1946 to begin a new life with his wife on eighty acres next to his uncle's farm. Elmer began with some pinto beans and a variety of other crops until the 1970s. Then he started specializing in hogs and ran a 250-sow operation. The hog business proved successful, but Elmer faced a long drive south to Phoenix to take his hogs to market and had no control over wholesale prices. As the surrounding county's population grew, Elmer seized the opportunity to diversify the products produced on the farm

and to market them directly to retail customers. By direct marketing, Elmer got a good price for his farm produce. Even more important, the community started coming out to see the farm. Now in his seventies, Elmer Young still works on the farm part-time. His son Gary and succeeding generations maintain the operation year-round.

From these humble beginnings, Young's Farm has flourished. The Youngs still raise animals – cows, chickens, and turkeys – but today they primarily grow produce: everything from corn to pumpkins, onions, cucumbers, and tomatoes. The Youngs practice careful stewardship of their land and water. They rotate crops, cycling the fields from red clover and beardless barley hay to sweet corn, then back again to hay, providing fallow time, reducing tillage, and restoring nitrogen and other key nutrients to the soil. Water conservation efforts include laser-leveling of their fields, planting only half the land at one time, and using surge and pump-back irrigation systems.

While farming is what they do best, the Youngs have also worked to transform themselves into the family farm of the future. They host nearly half a million visitors a year at their on-farm retail sales center and for ever-popular

The produce stand at Young's Farm is one of the many ways this family farm has managed to sustain itself.

weekend events, including sweet corn, pumpkin, garlic, and pie festivals.

Early on, the Youngs opened a roadside produce stand that still brims full of vegetables and fruit in summer and fall. Through the years, things have only gotten bigger and better. Most every morning the parking lot at the busy intersection of Highways 69 and 169 is packed with cars. Customers buy fresh-picked tomatoes at the produce stand, jellies and warm bread from the Farm Store, and collectables from the TreeHouse Gift Shop. The Farm Restaurant bustles with morning regulars enjoying a home-style breakfast.

There's also a blacksmith shop and a "critter corral," where visitors can see a family of goats – just for fun.

Today, Young's Farm has its own public outreach coordinator, Sarah Young Teskey. Sarah, in her twenties, has bright blue eyes and golden hair that belie her assured and articulate manner. Of late, she explains, her days have been preoccupied with a monumental task – saving Young's Farm.

Sarah attributes the success of Young's Farm to strong community support and to the quality of the land. Lonesome Valley, where the farm is located, once harbored a huge marshy lakebed created by a natural dam on the Agua Fria River. Water flowing down from the surrounding hills collected during wet periods and left behind sediment and debris that made rich soil. The unusually fertile ground is a precious resource worth saving. "We're adamant about preserving this for what it's good to grow," she insists. "This soil is so unique, especially for Arizona, that it just deserves the right to grow what it's supposed to be growing. It's not supposed to be covered in asphalt, in cement, in houses."

But such development is exactly what Sarah, her family, and her community are fighting against. The threat arose from a 1998 state decree that the water table of the Prescott Active Management Area must be balanced by the year 2025. Farms use more water than residences, and so a formula was devised that puts pressure on farmers to convert their irrigation usage to consumptive use. Each year they wait, the water associated with residential development of the land diminishes by a set percentage. Thus, it behooves a farmer to sell sooner rather than later.

To save their farm, the Youngs have joined with the Trust for Public Land and the Central Arizona Land Trust to raise $3.5 million to purchase the development rights for the farm. Under an agreement, the Trust would buy the development rights, the proceeds would go to the Youngs, and then a conservation easement would assure that the land remains undeveloped in perpetuity. The project has achieved some success through grants from the federal government's Farm and Ranch Land Protection Program, private foundations, and individuals. The Youngs also helped to pass the Arizona Agricultural Protection Act, which sets up a commission through which funds can be obtained for such projects, and they were instrumental in adding language to an Arizona law that will help secure the water supply for protected farmland. The legislature listened to the Youngs and wrote a ten-year reprieve from the water edict into state law.

Still, the Youngs face a major hurdle in raising all the money they need to match the federal grant. Many are keeping close watch on this cutting-edge case. For now, the future of this beloved part of the community tilts on uncertainty. But Elmer Young's granddaughter Sarah speaks with resolute reassurance in her voice: "My family loves this place, and we're going to keep doing what we love to do."

GOOD FRIENDS BUYING GOOD FOOD
The Flagstaff Community Market

Flagstaff, Arizona · By Peter Friederici

Vendors at the community market

On one sunny September morning Helen Fairley buys some green chilis from Charles van Riper to go along with the purple potatoes and yellow onions she's bought from Farmer Frank at the next stand. She's never had purple potatoes before but has heard they are creamy and sweet and well suited to potato salad.

Helen plans to roast the chilis, peel off their blackened skins, and serve their tangy flesh to friends with the potato salad. Charles and his wife, Sandra, grew the chilis in a raised bed on their Cedar Valley Ranch farm ten miles east of Flagstaff, fertilizing them with manure from the Merino sheep Sandra raises for wool. Their display table is opulent with peppers and sweet Japanese cucumbers and a single flawless kohlrabi plant that looks much too beautiful to eat. Bags of dried herbs and peppers are piled at one end. At the other Sandra, next to the $15 spindle-and-wool kits she sells, demonstrates to two shoppers how to make Merino yarn on a drop spindle. "Relax your fingers," she says, "then pull up gently." Just past her Dom Flemons, a

beanstalk of an itinerant musician with a thick mop of hair, is strumming his guitar and singing "the first cut is the deepest" – Saturday night lyrics transposed to Sunday morning.

"The farmers market is not only a great way to run into your friends, but also to support small farms," Helen is saying – "good friends buying good food." Then she has to break off. To the east a freight train rumbles toward the market. The engineer blows the train's piercing whistle. The regulars plug fingers into their ears. Conversations at the Flagstaff Community Market are often interrupted by the passage of Burlington Northern Santa Fe freights that roar by, some of them pulling tank cars transporting industrial quantities of midwestern corn syrup to food-processing plants somewhere.

At the market, which takes place every Sunday morning in summer and early fall in a downtown parking lot, there's no question where the food comes from. The vendors are from Sedona and Cottonwood, Camp Verde and Glendale. On average, according to a survey conducted by Northern Arizona University's Center for Sustainable Environments, they drive their produce 80.3 miles to the market. That means they have to get up early on Sunday mornings, but it also means that

Right: The colors of fresh market produce tantalize the senses.

GOOD FRIENDS BUYING GOOD FOOD

Sunday morning shoppers in Flagstaff enjoy fresh produce from throughout northern Arizona.

their wares are much fresher than what local residents buy in the supermarket. That produce is shipped an average of 731 miles, with a concomitant loss in freshness and increase in fossil fuel use.

The farmers market began in Flagstaff in the summer of 2001 and has quickly become a local tradition. "People here like knowing where their food comes from," says Kate O'Connor-Masse, an engaging woman who with her husband Mike runs the Chino Valley Farms stand, which sells only tomatoes – albeit up to twenty-five different varieties. "They like the variety, freshness, and flavor. We get people who only want one certain kind of tomato. It's like seeing your buddies every week. It's more than buying and selling, it's community-building."

The market's popularity is only enhanced by the fact that Flagstaff, high and dry at 7,000 feet, has a tough climate for growing vegetables – as long-time residents say, growing a single tomato to ripeness before an early frost comes is enough to qualify a backyard grower as a master gardener. "People at this elevation are appreciative of what we bring because they can't grow it here," says Charles Jordan, who with his wife Elanora runs the Bent River Ranch in Clarkdale, 3,000 feet lower in elevation, and among other products sells six different types of garlic at the market. Indeed, many of the vendors here do come from lower elevations. They don't sell only vegetables and herbs. Across from Charles and Elanora's stand, Molly Sedlmeier and her husband Felipe are selling a variety of fresh, hot tamales, including a popular kind filled with wild-caught sockeye salmon. Molly's grandmother, who taught her the family tamale recipe back in San Antonio, Texas, may not have foreseen that her corn masa would ever surround Alaskan fish, but the result is splendid.

Across the way Dick Tinlin is selling bags full of some of the 60,000 pounds of pecans he raises every year on his thirty acres down in Camp Verde. He planted his 1,400 trees twenty-five years ago. He has a PhD in hydrogeology and during his work career would stop in every two weeks to water the trees. Now retired from his day job, he's the self-described "Head Nut" of what has become a year-round operation. Some of the nuts he's selling here are raw, some are roasted in savory peanut oil, and some are sweetened with a haze of cinnamon; today, two hours into the three-hour market time, he's already out of the spicy and popular Cajun-style.

At other stands shoppers can buy shade-grown coffee from El Salvador; sample fruit bars sold by Katie Harris of the nearby Mountain Harvest grocery; select homemade jams and salsas made by Ernie and Joe Riley in Phoenix; and purchase heirloom seeds for their own gardening efforts from long-haired John Munk of Thunderfoot EarthWorks, who says his governing philosophy is that "if we own our own foods and seeds, we have more of our individual freedom."

Buyers and sellers alike profess to look forward to Sunday mornings. "The thing I feel best about is selling quality produce harvested the morning before," says Charles van Riper. "See the dirt on my hands? It's fresh, it's good, it's healthy. I know because I grew it." He turns to sell a couple of small but sweet green bell peppers to local resident Lisa Connor, who says her two-year-old daughter Kalyana eats them "like apples."

To the west another freight train rumbles toward the market. The engineer blows the train's piercing whistle. The regulars plug fingers into their ears.

At the Flagstaff Community Market, everyone in the family has a chance to pitch in.

PRIDE IN WHAT YOU DO
Ye Ol' Geezer Meat Shop

Moab, Utah · By Sue and Tony Norris

Rich Evans

The small town of Moab is a haven for hikers and lovers of the majestic canyon country wilderness of southern Utah. But natural beauty, recreation, and cute little shops are not the only reason people might stop. Approaching Moab from the south on Highway 191, you can't help but notice the catchy hand-painted figure of a jocular fellow holding a huge meat cleaver in one hand and a sign in the other. It reads: Ye Ol' Geezer Meat Shop.

This mom-and-pop enterprise is owned and run by Rich Evans and his wife Pat, and it is one of the few real butcher shops left that offers locally grown, naturally raised, dry-aged beef and fresh pork products. At the back of the building is another of Rich's enterprises. He is pastor of Friends in Christ Lutheran Church.

When you enter the shop, you see four old-style meat cases beautifully arranged and chock-full of succulent steaks, chops, roasts, and fresh-ground hamburger and sausage. The scent of smoked barbeque ribs fills the air. Dry sausages hang from the ceiling, fresh preserves sit atop the counters, and wooden barrels offer up delicious jerky. A busy clerk shouts out a friendly greeting, and you're on your way to culinary heaven. Born in eastern Colorado in 1949 and raised in Moab, Rich is possessed of an infectious belly laugh that punctuates his more serious statements. He and Pat established the business in Moab in 1998 when the church called him back from Idaho, where they also ran a butcher shop. "We thought we might be able to retire from this business, but that didn't happen," he jokes. Both have been in the meat business through their adult lives in "old-fashioned" shops and supermarkets. They have witnessed dramatic changes in the way meat is processed over the last three decades.

According to Rich, in the late 1970s plants began to break down the fresh carcasses themselves, immediately sealing the meat in plastic and boxing it for shipment to stores. Moab's local supermarket now receives its meat already cut and wrapped from a plant in Kansas. The box beef method eliminates the need for an individual butcher to spend time and intensive labor handling, aging, and "breaking" the carcass. The new method is more economical, to be sure, but natural flavor and tenderness are sacrificed – a sacrifice Rich Evans is not prepared to make. Also, commercial cattle are finished off in less time and fed with hormones and antibiotics. Very little beef is fed out to choice grade, and over half of that comes from Canada. This explains why beef prices jumped drastically in 2003 when a single animal in Canada was diagnosed with mad cow disease, likely a result of feeding animal byproducts.

The difference in current and traditional methods is "like day and night," observes Rich. In contrast to the "get it out there, stick it to them" approach of the corporate giants, he proclaims a deep need to "take pride in what I'm putting out for customers."

For that reason, Rich begins by buying his beef from an old friend who ranches nearby. They collaborate from beginning to end to produce a quality product. The cattle are grass-fed and finished out on corn, barley, and alfalfa for 90 to 120 days in a small feedlot. "The high protein in the barley and alfalfa builds muscle and prevents the animal from putting on too much back fat," says Rich. The process produces optimal marbling – the mixture of fat and lean in a piece of meat.

Pat Evans processes locally grown meat in the Moab shop.

Rich starts to work out a carcass two weeks after the animal is slaughtered at a facility in nearby Monticello. For maximum flavor and tenderness, the prime cuts are aged an additional four to six weeks at thirty-five degrees Fahrenheit. During this dry-aging process, Rich explains, the natural bacteria and enzymes in the meat work to break down muscle tissue, increasing tenderness and reducing moisture by 10 percent, and producing firmness and less loss during cooking. This is in sharp contrast to the practice of letting the meat soak in its own blood for the two weeks or so that it takes to ship a plastic Cryovac package from the plant to the consumer.

As Rich talks about the special needs of each cut, he reveals his expertise and personal love of good meat. Chuck, round steaks, and roasts need less aging, for example, and hamburger needs to be ground fresh daily. He is adamant about that. Whatever hamburger isn't sold in his store the day it is ground is supplied to a local restaurant.

Some of Ye Ol' Geezer's customers come from as far away as Salt Lake City, but his regular customers mostly are summer visitors and locals who can't pay a lot for meat, even though many are health conscious. His prices for prime cuts may necessarily be higher than the supermarket's, but he is competitive on the cheaper cuts. Rich no longer needs to advertise, because "word of mouth is the best advertisement anyway," he says. Occasionally, he'll donate some steaks to a football team for a fundraiser or cater a corporate barbeque. He is convinced that once people try dry-aged beef, they'll never turn back.

Some loyal customers are supplied through e-mail orders, but Rich does not sell beef wholesale because he would have to have a separate facility for processing. Expanding his business would require certification as organic, certification by the U.S. Department of Agriculture, or both, and those costs would be prohibitive for him at present. Even maintaining state certification can be expensive. Furthermore, if Rich were to cater to restaurants, he would also be stuck with a surplus of the lower-grade cuts.

Rich has arranged with Lamar Walker of Future Farmers of America to buy about 80 percent of his pork from 4-H students who raise hogs according to his specifications. This arrangement ensures him a naturally raised product and provides a year-round market for the future farmers. And he passes on to them his pride in a superior product and a determined work ethic. Sowing such ethical seeds in his community is important to this butcher/pastor.

Rich laments the fact that the skill in aging and working out a whole carcass is "a dying art" in this "fast-food society," despite the willingness of many ranchers to raise meat naturally and consumer desire for locally grown, unadulterated meat. Working as an old-fashioned butcher is just not something many people want to do nowadays.

"It's a tough business requiring a unique combination of having the skill and being a good businessman," Rich notes, "which I'm not, but my wife is, thank God." He continues, "It's cold, hard work and very few kids want that nowadays. A huge percentage of people in the meat-packing plants are from south of the border because they are the only ones willing to do the work."

In addition, industry regulations require continual capital investment, a formidable obstacle to the independent operator. In 2002, Rich and Pat updated their plant at a cost of $50,000. But Rich believes his reward is in offering a better quality at a fair price with good service.

Rich Evans is proud of what he and Pat have accomplished. "Pride in what you do is the number one thing," he says. Regrettably, that attitude probably does make him an "ol' geezer" in today's society.

A STUNNING SANCTUARY
Garland's Lodge Orchard

Oak Creek Canyon, Arizona · By Ashley Rood and Rose Houk

Tucked into the stunning redrock cliffs of northern Arizona's famed Oak Creek Canyon is a beautifully tended orchard, with a few venerable apple trees still bearing fruit after eighty years. Garland's Lodge Orchard is a collection of fruit trees carefully tended by Rob Lautze and Mario Valeruz, two men dedicated to organic agriculture.

Rob Lautze began working at the orchard in the mid-1970s, when Garland's Oak Creek Lodge came under the ownership of Gary and Mary Garland. The lodge is known for its rustic elegance and gourmet dining amid the gorgeous setting of Oak Creek Canyon. Part of the unique ambiance is owed to the organic orchard and garden, which not only beautify the grounds but also supply the restaurant with tomatoes, green beans, squash, and other fresh produce for meals. In turn, the lodge helps support the orchard.

Oak Creek has a long history of apple-raising. The Garland's property was originally homesteaded in 1910. The Purtyman family put in the irrigation ditch from Oak Creek to the land. Jess Howard, son of legendary settler Jesse "Bear" Howard, proved up the homestead, and the first apple trees were planted around 1920. In those days, apples were hauled by wagon up to Flagstaff and elsewhere in the young state of Arizona, and Oak Creek apples became a highly desired product.

Rob Lautze

That legacy is proudly carried on by the Garlands, but growing methods have changed. Instead of chemical sprays and fertilizers, Rob Lautze brought a dedication to the idea of improving every aspect of the orchard through organic farming methods. Since the 1980s, Rob has experimented with insect repellents, composting, and manual thinning and harvest. The reward is an abundant annual harvest of incredibly tasty and highly marketable apples. Among the 400 or so trees in the orchard, most are apple but there are also some almond, peach, apricot, and cherry trees, whose sweet blossoms herald springtime in the canyon each year.

The main reason the orchard's produce is so fine, Rob says, is the proverbial "location, location, location." The canyon's walls create a microclimate, with temperatures rarely exceeding 100 degrees Fahrenheit in summer. On winter nights, cold air can drain into the canyon bottom, but in this particular spot the air seems to flow through rather than

The orchard at Garland's Lodge enjoys a nearly ideal location for growing premium organic fruit.

settling in, and so frost damage to tender fruit blossoms hasn't been a big problem. The availability of irrigation water, gravity-fed from the original ditch off Oak Creek, is another key to the orchard's success. Garland's also enjoys a ready roadside market, with several million visitors driving Highway 89A through the scenic canyon each year.

While such attributes are surely beneficial, it is Rob's adept and passionate management that has ensured the quality of Garland's fruit. He knows the nuances of the orchard as intimately as he knows himself. On a walk through, he points out subtle details with a sparkle in his eyes. Each tree, branch, insect, and whim of weather earns his attention.

"Growing fruit," he adds, "is a full-time job if you do it seriously, especially if you grow organically."

On a sunny February morning, while the leafless trees were being hand-pruned in the lower orchard, Mario showed Rob a small white larva living under a piece of bark. It was a codling moth larva, public enemy number one of all apple growers. One of the biggest efforts at Garland's Orchard is the practice of integrated pest management. Rob has become an expert at using pheromone treatments to disrupt the mating of codling moths. The larvae overwinter under the bark, and the moths hatch about the time the trees bloom. Rob has adopted a two-pronged approach to managing both forms of the insect.

On the trees and strung on overhead wires are pheromone dispensers – about 400 of them per acre on the five acres of orchard. They mimic the scents that attract male moths to females, confusing the males and leaving them unable to find the females. With mating slowed down, the trees then are sprayed all over with a biological substance called *Bacillus thuringiensis*, or Bt. It's actually a special mix of Bt, molasses, brewer's yeast, and dried milk. For the eggs that do hatch, this "Bt mix" deters the worms from boring into the core and ruining apples. In addition, when the crew hand-thins the new apples in June, they remove any fruit that shows worm damage. "It's labor intensive," Rob admits, "but worth doing. Fruit damage at harvest has dropped to only about 5 to 10 percent."

With insect pests under control, soil fertility has become the big emphasis. After a dozen years of no chemicals, the soil has had time to cleanse itself. Along with generous applications of traditional compost, Rob has turned to compost tea. The "tea" is simply water mixed with rich compost in burlap bags, aerated, and left for a time to brew. This concoction is sprayed on the trees early in the season, at the "green tip" stage just before bloom. The tea acts as a fungicide against pests like powdery mildew, and provides an added nutrient boost. Soil tests have convinced Rob that it works better than commonly used sulfur additives, and he's greatly encouraged by the trees' improved health.

Garland's cultivates a diverse selection of apples, including heirloom varieties like old-fashioned Red Delicious, Grimes Golden, Jonathan, Stayman-Winesap, Lodi, Yellow Transparent, and Arkansas Black. In late July or early August,

buds from the old varieties are grafted onto "mother" rootstock in a process called t-budding. Heirloom varieties possess certain virtues. Arkansas Blacks, for example, were renowned for their keeping quality, Rob notes, and homesteaders could store them in the cellar. To him, these varieties just have "more character" and "a depth of flavor" that is absent in the limited modern varieties.

To ensure the quality of the harvest, all the fruit is tree-ripened and hand-picked. Starting with Gravensteins in early September, Rob and Mario pick at least six times through October, setting this orchard apart from larger operations that only pick twice a season. The apples are loaded manually into a sorter and individually separated into market, juice, or food-processing grades. The market apples are sold along the road near the lodge and at local farmers markets, and the juice-quality apples are processed as cider. Garland's makes one of the few unpasteurized ciders available for purchase.

Rob obviously has delved deeply into the topic of organic farming, and now he can boast years of firsthand experience. His fundamental conclusion is that "you pay for chemical use" in lost vigor. And while he acknowledges that what he's done at Garland's has been possible because of the small scale and the commitment of Gary and Mary Garland, he holds firm to the belief that it's worth every bit of the effort. In color, quality, and flavor, he simply knows the apples are better.

"Every year, as you stay in touch with what's happening in ecological and sustainable-oriented research, there are new issues. It keeps it an exciting and growing experience," he says. "I think it's a real coming thing."

FOOD AS MEDITATION
Hell's Backbone Grill

Boulder, Utah · By Rose Houk

*Jen Castle and
Blake Spalding*

At dusk on a summer evening, honey-warm light shines from the expanse of windows in the large timber and tin-roofed building. Kids tumble on the lawn out front, and horses graze in lush green grass in the pasture. While people dine in shorts and t-shirts on the patio, curious cats circle a small pond. Saucer-sized hollyhocks frame the stone path that leads up to the screen door, and Tibetan prayer flags wave in the sagebrush-scented air.

The place is the small town of Boulder, high on the Aquarius Plateau overlooking the slickrock wilderness of southeast Utah. The restaurant is Hell's Backbone Grill, named for the narrow nearby bridge that first linked Boulder to the outside world in the 1930s.

In many ways this restaurant is an unusual, and unexpected, surprise in this sparsely settled region. The owners are two women, Blake Spalding and Jen Castle.

Both were motivated to start the business by a strong commitment to providing delicious, healthful organic food with lots of heart and soul. "Our aim," they profess, "is to be a restaurant with a conscience."

For Blake, it's a matter of putting her Buddhist teachings into practice: "I've learned to meditate while cooking to infuse the food with a quality of loving-kindness and generosity." Every decision she and Jen make gets put through a "filter": is it a sustainable choice, is it good for the community, does it reflect their value system?

Both women were working as cooks when they met – Blake for river companies and Jen as a baker for a coffeehouse in Flagstaff, Arizona. They had adventures together in backcountry catering, and when the opportunity arose to buy the restaurant adjoining the rustic Boulder Mountain Lodge, they decided to take the plunge. "We had a similar pace and energy," Jen recalls, and an "intuitive taste" about food. And, she laughs, they knew the place was a real restaurant because it already "came with a sign."

The first two years proved tough in the tight-knit Mormon town, population 200, give or take a few. But as they became more comfortable in the community and business improved, things started to look up. National media attention hasn't hurt (the grill has received favorable notice in *Sunset* and *Oprah* magazines and *The New York Times*, to name a few). After four years of very hard work, both Jen and Blake smile and say they've realized a dream come true in a place they consider paradise.

At Hell's Backbone Grill, every dish is made with fresh, seasonal, and, when possible, organic and locally produced ingredients. Both Blake and Jen design all the menus. A spring dinner will feature asparagus soup, a summer breakfast may offer fresh cherry pancakes, a fall meal will end with pear gingerbread. Other common ingredients are corn, piñon nuts, goat cheese, and juniper berries for seasoning. Much of this reflects Jen's New Mexico upbringing in a big family, where there was "always a lot of cooking, canning, and gardening." Basically, Blake notes, on any given day they look at what they have that they can turn into a dish.

With the nearest "real" grocery store a couple hours away, creativity and resourcefulness are requirements. Blake and Jen do receive regular food deliveries from Salt Lake City and a California co-op, but pride themselves on purchasing locally whenever they can. Beef and lamb come from neighboring ranchers, and trout from a fish farm in a nearby town. Organic chicken is harder to come by, but is on the menu at times. Again, diners at Hell's Backbone can rest assured that any meats have been selected from animals that were well fed and well cared for. That is a Mormon tradition anyway, Blake observes, and so their neighbors readily understand what they want. Local eggs have huge yolks that

The chefs at Hell's Backbone Grill pride themselves on creative and seasonal dishes.

are an "unearthly color of orange," she adds. An employee and longtime local resident makes tortillas and a special "she-devil sauce" for those brave enough to try it. Another specialty is Jen's Lemon Chiffon Cake, a blue-ribbon winner at the county fair.

REGIONAL FOOD HERITAGE

Jen and Blake employ a full-time gardener who harvests innumerable vegetables – lettuce, carrots, cucumbers, heirloom tomatoes, chiles, squash, beets, beans, peas – along with strawberries, rhubarb, and many herbs. A nearby orchard overflows with heirloom apples, peaches, pears, apricots, and small yellow plums. "In summer, whatever we have going on with fruits, you'll see it on the menu," remarks Blake – plum butter, plum glaze, plum jam, even plum barbeque sauce. They also raise beautiful old-fashioned flowers – pansies, nasturtiums, marigolds, zinnias, and daisies. Many are edible; a steak may come served with snips of marigold petals, or fresh limeade is garnished with a demure violet.

Boulder is situated at 6,800 feet in elevation, and the Aquarius Plateau is blessed with winter snow, summer rain, and good growing conditions. It would be more convenient just to order something from "somewhere far away," says Blake, "but food that's prepared and eaten in the place where it is also produced has the very specific and special flavor of the land. That's what travelers to our restaurant really seem to respond to."

Their thoughtful decision-making is evident not only in the food but in every other detail, including the handling of trash. The town of Boulder does not have a recycling center or even curbside garbage pickup. The rule is that "everything we grow, buy, and order must be used and used and reused," vow Jen and Blake. For example, the restaurant reduces waste by serving only draft beer, rather than beer in glass bottles.

Composting is a given. Table scraps are collected to feed local animals.

With tables filled nearly every night during the season from mid-March to the end of October, the restaurant has been a great success. It seats about sixty diners at one time, but when asked how many they serve each week or month, Blake and Jen don't have a ready answer. "We pay more attention to personal connections than numbers," replies Blake. She and Jen have lately been discussing what they might want to do should they have money left over. And, not surprisingly, they would like to put it to good use locally too: "build a park, buy violins for the school kids, pay higher wages," notes Jen. They've also started a yearly tradition of a musical Folkfest and Farmers Market in August that features local talent and wares.

In the fall of 2004, Blake and Jen self-published their own cookbook, *With a Measure of Grace: The Story and Recipes of a Small-Town Restaurant.* Co-written with Blake's sister, Lavinia Spalding, and photographed by Eric Swanson, it contains "signature recipes, lively anecdotes, and a healthy dose of old-fashioned rural wisdom."

And now that they've "made it," so to speak, Blake has found that the most rewarding part is "promoting an 'earth-to-table' eating experience." For Jen, it's knowing that she's preparing and serving food that's "really honest and really true."

Left: Hell's Backbone Grill lights up the culinary landscape of southern Utah.

A New Plateau

ACKNOWLEDGMENTS

This book was primarily funded by a Ford Foundation grant to the Center for Sustainable Environments; thanks to former program officers Ken Wilson and Damien Pwono for their support. Additional financial support for fieldwork on particular topics in this book was provided by the Arizona Proposition 301 ERDENE initiative; Agnese Haury; the C.S. Fund; the Greenville Foundation; the Compton Foundation; the Laird Norton Endowment Foundation; the Town Creek Foundation; the John and Sophie Ottens Foundation; and the CSE Director's Fund. Project advisors for the CSE's Ford Foundation project included Tim Richard, Warren Miller, Sue Bruckner, Arnold Taylor, Micah Lomoamvaya, John Blueyes, and Kim Howell-Costion.

The book's core editorial team in Flagstaff consisted of Peter Friederici, Rose Houk, Gary Paul Nabhan, Roger Clark, and Tony Norris. Other writers included Sue Norris, Susan Lamb, David Seibert, and Rachel Turiel Hinds. Much of the fieldwork for the book was conducted and documented by CSE interns Tim Swinehart, Rosemary Logan, Patty Kohany, Ben Bellarado, Roberto Nutlouis, Lilian Hill, Ashley Rood, Rebecca de Groot, Charles E. Jones, and Charlie Laurel.

Many talented photographers also contributed to the project. Tony Marinella of the Museum of Northern Arizona traveled many miles in pursuit of compelling images, and was indispensable in the process of acquiring and organizing the photographs.

This project was completed in partnership with Renewing the Countryside, who received generous support from the W. K. Kellogg Foundation and USDA – Cooperative State Research, Education, and Extension Service. The Renewing the Countryside team included Brett Olson, art director and designer for the book, Jan Joannides, series editor, and Hannah Lewis who contributed to the editing process.

May readers draw as much inspiration from these pages as we did in putting them together.

A New Plateau

—

PUBLISHING PARTNERS

Center for Sustainable Environments

The Center for Sustainable Environments at Northern Arizona University is a catalyst for collaborative research, education, training, and stewardship among its diverse partners. It brings together the talents and expertise of scientists, educators, independent scholars, business leaders, government agencies, nonprofit organizations, students, and community members to seek creative solutions to environmental problems. These challenges are addressed through initiatives that safeguard natural and cultural values and resources. By combining technical innovations with the knowledge, values, and practices of local communities, CSE generates long-term environmental solutions that enhance the lives of those they impact. Read about CSE's current projects at www.environment.nau.edu.

Museum of Northern Arizona

The mission of the Museum of Northern Arizona is to inspire a sense of love and responsibility for the beauty and diversity of the Colorado Plateau through collecting, studying, interpreting, and preserving the region's natural and cultural heritage. Located in Flagstaff, Arizona, the Museum is the gateway to understanding the Grand Canyon region's rich cultural and ecological diversity. It offers changing exhibits, informative programs, and outdoor adventures to encourage a deeper knowledge of the area's unique beauty and character. Its educational programs engage people of all ages in experiencing artistic traditions, Native cultures, and natural history through extended expeditions across the Four Corners states of Utah, Colorado, New Mexico, and Arizona. The Museum of Northern Arizona has proudly served as this region's ambassador and author of its place-based curricula since it was founded in 1928. More information is available at www.musnaz.org.

STORY CONTACTS

1. **Stone Free Farm**
   Chuck Barry and Rosie Carter
   Cortez, CO
   rcarter@gobrainstorm.net

2. **Lost Cabin Ranch**
   Rebecca Routson
   Prescott, AZ

3. **Ferrell Secakuku**
   Second Mesa, AZ

4. **Sutcliffe Vineyards and McElmo Canyon Wines**
   John and Emily Sutcliffe
   Cortez, CO
   970-565-0825
   battlerock@frontier.net

5. **Diné, Inc.**
   Hank Willie
   P.O. Box 7087
   Winslow, AZ 86047
   928-657-3272
   navatec@starband.net

6. **Ashokola Gardens**
   Kim and Joseph Costion
   P.O. Box 504
   Snowflake, AZ 85937
   ashokalagardens@aol.com

7. **Vernon Masayesva**
   P.O. Box 33
   Kykotsmovi, AZ 86039
   928-734-9255
   kuuyi@aol.com
   www.blackmesatrust.org

8. **Oakhaven Permaculture Center**
   Tom Riesing and Chris Berven
   4179 County Road 124
   Hesperus, CO 81326
   970-259-5445
   triesing@oakhavenpc.org
   cberven@oakhavenpc.org
   www.oakhavenpc.org

9. **Whipstone Farm**
   Cory Rade and Shanti Leinow
   21640 North Juniper Ridge Road
   Paulden, AZ 86334
   928-636-6209
   whipstone@aol.com

10. **Shared Harvest Community Garden**
    Bob Kauer
    3232 County Road 234
    Durango, CO 81301
    970-247-7850

11. **Diné Be'iiná/Sheep Is Life**
    Roy Kady
    P.O. Box 209
    Teec Nos Pos, AZ 86514
    928-656-3498
    roykady@navajolifeway.org
    www.dinewoven.com

12. **Shepherd's Lamb**
    Antonio and Molly Manzanares
    P.O. Box 307
    Tierra Amarilla, NM 87575
    505-588-7792
    shepherd@rioarriba.com
    www.organiclamb.com

13. **Babbitt Ranches**
    Billy Cordasco
    P.O. Box 520
    Flagstaff, AZ 86002-0502
    928-774-6199
    cobar@babbittranches.com
    www.babbittranches.com

14. **Stargate Valley Farms**
Carol Poore and Dennis Swayda
P.O. Box 97
Woodruff, AZ 85942
928-524-7717
www.stargatevalleyfarms.com

15. **Diablo Trust**
P.O. Box 3058
Flagstaff, AZ 86003
928-523-0588
info@diablotrust.org
www.diablotrust.org

16. **Winter Sun Herbals**
Phyllis Hogan
107 North San Francisco Street
Flagstaff, AZ 86001
928-774-2884
WSAERA@hotmail.com
www.wintersun.com

17. **Turtle Lake Refuge**
Katrina Blair
848 East 3rd Avenue
Durango, CO 81301
970-247-8395
turtlelakerefuge@yahoo.com
www.turtlelakerefuge.org

18. **Basket Makers of the San Juan Paiute**
Mabel Lehi
P.O. Box 985
Tuba City, AZ 86045
928-283-5536

Clyde Whiskers
P.O. Box 10111
Tonalea, AZ 86044
928-606-4766

19. **Zuni Furniture Enterprise**
Sterling Tipton
164 Route 301 North (Black Rock)
Zuni, NM 87327
505-782-5855
zfe@newmexico.net
www.ashiwi.org/enterprises/enterprise.htm

20. **Melissa Porter**
P.O. Box 394
Chama, NM 87520
505-756-2492
porters@cvn.com

21. **Mountain Top Honey**
Dennis Arp
383 Choctaw
Flagstaff, AZ 86001
928-525-1671

22. **Community Wild Foraging Project**
Patty West
Center for Sustainable Environments
Northern Arizona University
P.O. Box 5765
Flagstaff, AZ 86011-5765
928-523-2942
Patty.West@nau.edu
www.environment.nau.edu

23. **Solar Design & Construction**
Ed Dunn
21 West Pine Avenue
Flagstaff, AZ 86001
928-774-6308
solarbale@cybertrails.com

24. **Sustainable Housing for Indigenous People Project**
Lilian Hill
P.O. Box 79
Kykotsmovi, AZ 86039
928-734-9426
topetqoatl@yahoo.com

25. **w/Gaia Design**
Susie Harrington
P.O. Box 264
Moab, UT 84532
susie@withgaia.com
www.withgaia.com

26. Charles A. Laurel,
    Building Consult & Design
    Charlie Laurel and Margaret Spilker
    P.O. Box 30296
    Flagstaff, AZ 86003
    calaurel@hotmail.com

27. Indigenous Community Enterprises
    2717 N. Steves Boulevard, Suite 8
    Flagstaff, AZ 86004
    928-522-6162
    www.cba.nau.edu/ice/

    SouthWest Tradition Log Homes
    Ron Taylor
    P.O. Box 468
    Cameron, AZ 86020
    928-679-2031
    roundwud@hotmail.com

28. A.C.E. Builders, Inc.
    Jack Ehrhardt
    2170 Northern Avenue #B
    Kingman, AZ 86401
    928-757-4202
    cerbatnp@citlink.net
    www.ecolifetours.com

29. Watson Conserves
    Mac Watson
    1517 Canyon Road
    Santa Fe, NM 87501
    macwatson@cybermesa.com

30. Native Sun
    Doran Dalton
    P.O. Box 660
    Kykotsmovi, AZ 86039
    info@nativesun.biz
    www.nativesun.biz

31. Southwest Windpower
    Andy Kruse
    2131 North First Street
    Flagstaff, AZ 86004
    928-779-9463
    info@windenergy.com
    www.windenergy.com

32. Turquoise Room
    John Sharpe
    303 East Second Street
    Winslow, AZ 86047
    928-289-2888
    turquoise@winslow-az.net
    www.laposada.org/restaurant.htm

33. Cow Canyon Trading Post
    Liza Doran
    P.O. Box 88
    Bluff, UT 84512
    435-672-2208

34. Young's Farm
    Sarah Teskey
    P.O. Box 147
    Dewey, AZ 86327
    928-632-7272
    info@youngsfarminc.com
    www.youngsfarminc.com

35. Flagstaff Community Market
    Flagstaff, AZ

36. Ye Ol' Geezer Meat Shop
    Rich and Pat Evans
    1240 South Highway 191
    Moab, UT 84532
    435-259-4378
    geezer_meats@citlink.net
    www.nav.to/olgeezer

37. Garland's Oak Creek Lodge and
    Orchard
    Rob Lautze
    1055 Shady Lane
    Sedona, AZ 86336
    928-282-1723

38. Hell's Backbone Grill
    Blake Spalding and Jen Castle
    P.O. Box 1428
    Boulder, UT 84716
    435-335-7464
    hellsbackbonegrill@color-country.net
    www.hellsbackbonegrill.com

A New Plateau

PHOTOGRAPHY CREDITS

RENEWING THE COUNTRYSIDE

Renewing the Countryside (RTC) is a non-profit organization that publishes books, calendars, and a website that shares stories of people who are redefining what it means to live, work, and learn in rural America. RTC finds stories of rural renewal through partnerships with other organizations that not only have similar goals of strengthening rural communities and the environment, but also bring specific expertise within a region or topic area. RTC works in partnership with these organizations to champion rural communities, farmers, ranchers, artists, entrepreneurs, educators, activists, and other rural heroes.

While there are an abundance of stories of innovative entrepreneurs and creative community initiatives in rural areas, these stories rarely are seen in the mainstream media and almost never in the bindings of a coffee-table style book. This format gives legitimacy and validity to this new vibrancy and sustainability in the countryside. RTC's publications expose a new group of people to a side of the rural landscape they may otherwise never hear about. They also serve as a resource guide for people to examine their own rural enterprises and help them craft more sustainable and vibrant communities.

Through the generous support of many individuals and foundations, Renewing the Countryside is dedicated to sharing the strength of America's rural landscape: the people enhancing their cultural and natural resources while spurring local economic development in their communities.

www.renewingthecountryside.org

Renewing the Countryside · 2105 First Avenue South · Minneapolis, Minnesota 55404 USA · 1-866-378-0587 · info@rtcinfo.org